Conference Proceedings of the Society for Experimental Mechanics Series

Series Editor

Kristin B. Zimmerman, Ph.D.
Society for Experimental Mechanics, Inc.,
Bethel, CT, USA

The Conference Proceedings of the Society for Experimental Mechanics Series presents early findings and case studies from a wide range of fundamental and applied work across the broad range of fields that comprise Experimental Mechanics. Series volumes follow the principle tracks or focus topics featured in each of the Society's two annual conferences: IMAC, A Conference and Exposition on Structural Dynamics, and the Society's Annual Conference & Exposition and will address critical areas of interest to researchers and design engineers working in all areas of Structural Dynamics, Solid Mechanics and Materials Research.

More information about this series at http://www.springer.com/series/8922

Kirk Grimmelsman
Editor

Dynamics of Civil Structures, Volume 2

Proceedings of the 39th IMAC, A Conference and Exposition on Structural Dynamics 2021

Springer

Editor
Kirk Grimmelsman
Intellectual Infrastructure Systems
Philadelphia, PA, USA

ISSN 2191-5644 ISSN 2191-5652 (electronic)
Conference Proceedings of the Society for Experimental Mechanics Series
ISBN 978-3-030-77142-3 ISBN 978-3-030-77143-0 (eBook)
https://doi.org/10.1007/978-3-030-77143-0

This Springer imprint is published by the registered company Springer Nature Switzerland AG
The registered company address is: Gewerbestrasse 11, 6330 Cham, Switzerland

Preface

Dynamics of Civil Structures represents one of the nine volumes of technical papers presented at the 39th IMAC, A Conference and Exposition on Structural Dynamics, organized by the Society for Experimental Mechanics and held on February 8–11, 2021. The full proceedings also include volumes on Nonlinear Structures and Systems; Model Validation and Uncertainty Quantification; Dynamic Substructures; Special Topics in Structural Dynamics and Experimental Techniques; Rotating Machinery, Optical Methods and Scanning LDV Methods; Sensors and Instrumentation, Aircraft/Aerospace, Energy Harvesting and Dynamic Environments Testing; Topics in Modal Analysis and Parameter Identification; and Data Science in Engineering.

Each collection presents early findings from analytical, experimental, and computational investigations on an important area within structural dynamics. Dynamics of civil structures is one of these areas which cover topics of interest of several disciplines in engineering and science.

The Dynamics of Civil Structures Technical Division serves as a primary focal point within the SEM umbrella for technical activities devoted to civil structures analysis, testing, monitoring, and assessment. This volume covers a variety of topics, including structural vibrations, damage identification, human–structure interaction, vibration control, model updating, modal analysis of in-service structures, innovative measurement techniques and mobile sensing, and bridge dynamics among many other topics.

Papers cover testing and analysis of different kinds of civil engineering structures, such as buildings, bridges, stadiums, dams, and others.

The organizers would like to thank the authors, presenters, session organizers, and session chairs for their participation in this track.

Philadelphia, PA, USA — Kirk Grimmelsman

Contents

Particle Filters and Auto-Encoders Combination for Damage Diagnosis on Hysteretic Non-Linear Structures Subject to Changing Environmental Conditions 1
Luca Lomazzi, Francesco Cadini, and Marco Giglio

A New Benchmark Problem for Structural Damage Detection: Bolt Loosening Tests on a Large-Scale Laboratory Structure 15
Onur Avci, Osama Abdeljaber, Serkan Kiranyaz, Mohammed Hussein, Moncef Gabbouj, and Daniel Inman

Implementation of an Organic Database Structure for Population-Based Structural Health Monitoring 23
Daniel S. Brennan, Chandula T. Wickramarachchi, Elizabeth J. Cross, and Keith Worden

Estimation of Blade Forces in Wind Turbines Using Strain Measurements Collected on Blades 43
Bridget Moynihan, Babak Moaveni, Sauro Liberatore, and Eric Hines

On Health-State Transition Models for Risk-Based Structural Health Monitoring 49
A. J. Hughes, R. J. Barthorpe, and K. Worden

Cointegration for Structural Damage Detection Under Environmental Variabilities: An Experimental Study 61
J. C. Burgos, B. A. Qadri, and M. D. Ulriksen

Footbridge Vibrations and Modelling of Pedestrian Loads 69
Lars Pedersen and Christian Frier

Multi-LSTM-Based Framework for Ambient Intelligence 75
Nur Sila Gulgec, Martin Takáč, and Shamim N. Pakzad

Operational Modal Analysis and Finite Element Model Updating of a 53-Story Building 83
Onur Avci, Khalid Alkhamis, Osama Abdeljaber, and Mohammed Hussein

An Overview of Deep Learning Methods Used in Vibration-Based Damage Detection in Civil Engineering 93
Onur Avci, Osama Abdeljaber, and Serkan Kiranyaz

Transfer Learning from Audio Domains a Valuable Tool for Structural Health Monitoring 99
Eleonora M. Tronci, Homayoon Beigi, Maria Q. Feng, and Raimondo Betti

Experimental Evaluation of Drive-by Health Monitoring on a Short-Span Bridge Using OMA Techniques 109
William Locke, Laura Redmond, and Matthias Schmid

Investigation of Low-Cost Accelerometer Performance for Vibration Analysis of Bridges 129
Kirk Grimmelsman

Real-Time Human Cognition of Nearby Vibrations Using Augmented Reality 139
Elijah Wyckoff, Marlan Ball, and Fernando Moreu

Understanding Errors from Multi-Input-Multi-Output (MIMO) Testing of a Cantilever Beam 147
Fernando Moreu, James Woodall, and Arup Maji

Load-Displacement Behavior Clustering of RC Shear Walls Using Functional Data Analysis 153
Hamed Momeni and Arvin Ebrahimkhanlou

Preface

Dynamics of Civil Structures represents one of the nine volumes of technical papers presented at the 39th IMAC, A Conference and Exposition on Structural Dynamics, organized by the Society for Experimental Mechanics and held on February 8–11, 2021. The full proceedings also include volumes on Nonlinear Structures and Systems; Model Validation and Uncertainty Quantification; Dynamic Substructures; Special Topics in Structural Dynamics and Experimental Techniques; Rotating Machinery, Optical Methods and Scanning LDV Methods; Sensors and Instrumentation, Aircraft/Aerospace, Energy Harvesting and Dynamic Environments Testing; Topics in Modal Analysis and Parameter Identification; and Data Science in Engineering.

Each collection presents early findings from analytical, experimental, and computational investigations on an important area within structural dynamics. Dynamics of civil structures is one of these areas which cover topics of interest of several disciplines in engineering and science.

The Dynamics of Civil Structures Technical Division serves as a primary focal point within the SEM umbrella for technical activities devoted to civil structures analysis, testing, monitoring, and assessment. This volume covers a variety of topics, including structural vibrations, damage identification, human–structure interaction, vibration control, model updating, modal analysis of in-service structures, innovative measurement techniques and mobile sensing, and bridge dynamics among many other topics.

Papers cover testing and analysis of different kinds of civil engineering structures, such as buildings, bridges, stadiums, dams, and others.

The organizers would like to thank the authors, presenters, session organizers, and session chairs for their participation in this track.

Philadelphia, PA, USA — Kirk Grimmelsman

Contents

Particle Filters and Auto-Encoders Combination for Damage Diagnosis on Hysteretic Non-Linear Structures Subject to Changing Environmental Conditions 1
Luca Lomazzi, Francesco Cadini, and Marco Giglio

A New Benchmark Problem for Structural Damage Detection: Bolt Loosening Tests on a Large-Scale Laboratory Structure 15
Onur Avci, Osama Abdeljaber, Serkan Kiranyaz, Mohammed Hussein, Moncef Gabbouj, and Daniel Inman

Implementation of an Organic Database Structure for Population-Based Structural Health Monitoring 23
Daniel S. Brennan, Chandula T. Wickramarachchi, Elizabeth J. Cross, and Keith Worden

Estimation of Blade Forces in Wind Turbines Using Strain Measurements Collected on Blades 43
Bridget Moynihan, Babak Moaveni, Sauro Liberatore, and Eric Hines

On Health-State Transition Models for Risk-Based Structural Health Monitoring 49
A. J. Hughes, R. J. Barthorpe, and K. Worden

Cointegration for Structural Damage Detection Under Environmental Variabilities: An Experimental Study 61
J. C. Burgos, B. A. Qadri, and M. D. Ulriksen

Footbridge Vibrations and Modelling of Pedestrian Loads 69
Lars Pedersen and Christian Frier

Multi-LSTM-Based Framework for Ambient Intelligence 75
Nur Sila Gulgec, Martin Takáč, and Shamim N. Pakzad

Operational Modal Analysis and Finite Element Model Updating of a 53-Story Building 83
Onur Avci, Khalid Alkhamis, Osama Abdeljaber, and Mohammed Hussein

An Overview of Deep Learning Methods Used in Vibration-Based Damage Detection in Civil Engineering 93
Onur Avci, Osama Abdeljaber, and Serkan Kiranyaz

Transfer Learning from Audio Domains a Valuable Tool for Structural Health Monitoring 99
Eleonora M. Tronci, Homayoon Beigi, Maria Q. Feng, and Raimondo Betti

Experimental Evaluation of Drive-by Health Monitoring on a Short-Span Bridge Using OMA Techniques 109
William Locke, Laura Redmond, and Matthias Schmid

Investigation of Low-Cost Accelerometer Performance for Vibration Analysis of Bridges 129
Kirk Grimmelsman

Real-Time Human Cognition of Nearby Vibrations Using Augmented Reality 139
Elijah Wyckoff, Marlan Ball, and Fernando Moreu

Understanding Errors from Multi-Input-Multi-Output (MIMO) Testing of a Cantilever Beam 147
Fernando Moreu, James Woodall, and Arup Maji

Load-Displacement Behavior Clustering of RC Shear Walls Using Functional Data Analysis 153
Hamed Momeni and Arvin Ebrahimkhanlou

Particle Filters and Auto-Encoders Combination for Damage Diagnosis on Hysteretic Non-Linear Structures Subject to Changing Environmental Conditions

Luca Lomazzi, Francesco Cadini, and Marco Giglio

Abstract Damages may naturally arise in structures within their life span due to the insurgence of phenomena related to normal operation. Their occurrence might also be favored by external boundary conditions the systems experience during their lifetime, such as time-varying environmental and operating conditions. Standard maintenance activities, such as scheduled non-destructive testing (NDT) and corrective maintenance, are typically carried out to improve the health and longevity of such systems, typically entailing long downtimes with significant economic impacts. In recent decades, condition-based maintenance strategies (CBM) or even predictive ones (PM) have increasingly gained popularity since, in principle, they allow to optimally intervene on the structure only when really required by its current conditions. These maintenance schemes require that a deep knowledge of the system current state of health and, possibly, of the main degradation mechanisms be available, which may rely on advanced structural health monitoring (SHM) systems being installed on the structures for performing real-time diagnosis and prognosis. Many approaches to SHM have been formulated, with several applications to mechanical, aeronautical, space, and civil structures. Particle Filters (PFs) have been proposed as a model-based, time-domain tool for estimating hidden, not observable system states, including those normally affected by damage, in particular, when the structure behavior is non-linear and affected by non-Gaussian disturbances and noises. Yet, in case of varying operating and environmental conditions, the SHM task often still turns out to be quite challenging, since the diagnostic features associated with damage can be significantly distorted. To overcome this issue, auto-encoders have successfully been employed to extract damage-related features in presence of such varying external conditions. Thus, this work aims at combining these two methods for developing an original approach to damage detection and localization in structures, robust with respect to changing environmental and operating conditions, capable of leveraging the specific benefits provided by the two aforementioned methodologies. The proposed algorithm is demonstrated with reference to the problem of damage diagnosis on a vibrating n-degrees of freedom system, featuring a non-linear stiffness component characterized by a Bouc-Wen hysteretic behavior and subject to varying temperature conditions.

Keywords Structural health monitoring · Damage diagnosis · Changing environmental conditions · Non-linear · Bouc-Wen

1 Introduction

Damage may arise within the lifetime of structural systems due to several favorable conditions such as occasional loads, time-varying environmental and operating conditions, and long service times. That is the case, for example, of civil structures, the damages of which are usually determined by aging [1] and by the constantly increasing loads they are exposed to, e.g., the frequency and the mass of vehicles crossing bridges. Those effects are also experienced by mechanical systems, which are as important to society as civil structures. For instance, that is common to wind turbines [2], whose 20 years designed service life in most of the cases has already been reached or will be in the nearest future in Europe. In the general framework of structures, several damages may occur, such as fatigue, creep, erosion, or corrosion. Those may arise and accumulate within the service life not only because of extreme events, such as generic overloads from earthquakes and winds, but also due to aging, which affects structures in normal operating conditions, eventually determining failure conditions.

L. Lomazzi · F. Cadini (✉) · M. Giglio
Dipartimento di Meccanica, Politecnico di Milano, Milano, Italy
e-mail: luca.lomazzi@polimi.it; francesco.cadini@polimi.it; marco.giglio@polimi.it

K. Grimmelsman (ed.), *Dynamics of Civil Structures, Volume 2*, Conference Proceedings of the Society for Experimental Mechanics Series, https://doi.org/10.1007/978-3-030-77143-0_1

In this outlined context, among the many effective ways to increase the return on investment of the existing assets [3], lengthening the lifetime of such systems turns out to be particularly appealing. To this purpose, novel maintenance policies, such as condition-based and predictive maintenance techniques, are constantly gaining importance over time. Those rely on the information automatically gathered and processed by a net of sensors embedded in structures, performing the so-called Structural Health Monitoring (SHM). A quite ample literature is available on time and frequency domain-based methods that may be exploited to identify possible damage-induced deviations of features extracted from measured data, with the purpose of performing damage detection, localization, quantification, and, eventually, prognosis. Envelope spectrum [4], cepstrum [5], high-order spectrum, and coherence function are among the frequency domain-based methods [6]. Moreover, the exploitation of transmissibility functions (TFs) for performing damage detection is worth a particular mention [7, 8]. In fact, those both do not require any prior knowledge of the structural model and have been proved to always be damage sensitive [8]. Additionally, TFs may be exploited to perform a further step of the SHM procedure, i.e., damage localization [9], even though criticalities may arise in this practice [10]. On the other hand, time-waveform indices, orbits, probability density functions, and probability density moments are typical time domain-based methods employed in the SHM framework. By applying such methods a huge amount of data is accumulated over the lifetime of the monitored structure, with both the advantage of leading to detailed damage analyses, and the disadvantage of increasing the required resources for performing damage identification at any complexity level [6].

However, all the classical feature extraction methods are unable to distinguish between damage-driven deviations from the healthy structural behavior and those determined by a normal change in environmental and operating conditions, such as temperature and operative loads variations, thus sometimes leading to false damage detection alarms. This topic is widely discussed in [11, 12] and [13]. With the purpose of solving that issue, many methods known as data normalization [11] have been proposed in the available literature, which aim at suppressing the effects of the aforementioned confounding, yet inevitable, factors. Among those proposed approaches, singular-value decomposition [14], principal component analysis [15], factor analysis [16], cointegration [17, 18], and auto-associative neural networks [19] leverage on some identifiable feature shifts induced by damage, while other methodologies exist which exploit regression analysis to automatically detect the relationship between varying boundary conditions and measured features [20–23]. However, although possibly able to subtract the confounding factors effects from damage-related features, those proposed methods may only detect the presence of a damage, without further analyzing the data to localize and quantify it. An exception may be identified in the work by Limongelli [24], which proposes an interpolation-based damage detection method, leading, however, to frequent false alarms. Additionally, these methods are all completely tailored to a particular setup of interest, thus not allowing the generalization of the proposed approach to the damage assessment in generic structures, with few exceptions, e.g., [25].

More advanced time-domain methods employing the Bayesian framework have been recently exploited to perform structural parameters identification [26] and fault diagnosis [27]. Among those ones, the use of particle filters (PFs) seems to be particularly promising, since those can also cope with non-linear dynamics of the investigated system. For instance, PFs have been exploited in system parameters identification applications, such as the stiffness value identification in case of multiple-degree of freedom (MDOF) systems experiencing seismic-like accelerations [28]. However, to the best of the authors' knowledge, no exploitation of PFs for assessing structures in time-varying external conditions have been investigated yet for performing damage diagnosis. The PF potentiality of predicting the system dynamic behavior in presence of non-linear effects is particularly interesting in the civil engineering field. In fact, some civil structures, such as Reinforced Concrete (RC) frames, may show a non-linear behavior, in form of hysteretic behavior, when subject to dynamic loads, such as earthquakes. Many models have been proposed in the literature for accounting for the observed hysteretic behavior of such structures, (e.g., [29–31]). Among those many theories, the most reasonable and popular one is the Bouc-Wen model [32], which is widely adopted in parameters identification-aimed works [33].

Recently, damage diagnosis has been performed by means of neural networks [34, 35], thanks to the great capability of those machine learning tools of not requiring any physics-based model of the event under investigation, thus leading to damage-related features extraction even in case of complex dynamics. In particular, it seems particularly promising the exploitation of auto-encoders for performing damage detection [34, 36, 37]. For instance, those have been implemented in a frequency domain-based framework for automatically processing the information extracted from transmissibility functions built from vibration measurements [38], allowing to perform novelty detection in MDOF-modeled systems. However, as already mentioned, TFs may provide inaccurate results when employed for performing damage localization [39].

In this work, PFs and auto-encoders are combined into an original methodological approach, working in the time-domain, aimed at performing damage detection and localization in MDOF-like structures subject to changing environmental and external conditions, eventually characterized by non-linear dynamics. In particular, processing the information of available vibration measurements, i.e., accelerations, the PF provides an estimate of some not directly observable structural parameters. Auto-encoders are then employed for combining PF posterior estimation and temperature measurements, with the purpose of subtracting the effects of confounding influences from the damage-related features extraction process. Additionally, an automatic statistical threshold is implemented in the proposed algorithm as human-free decision-making tool for identifying the presence of a damage with a certain confidence level. The method is demonstrated with reference to a MDOF system whose parameters are taken from a literature benchmark case study [28]. The case study proposed in this work focuses on performing damage detection and localization considering a structure with a component affected by hysteretic behavior, with the aim of showing the proposed framework performances in presence of eventual non-linearities, which may severely hamper the damage diagnosis task.

The paper is organized as follows. In Sect. 2 the generic methodological approach is presented, focusing on the specialization of the selected tools to the MDOF system model. Section 3 presents a case study consisting of a 3-DOF system affected by temperature variations, experiencing an earthquake-like acceleration. In Sect. 4 a critical discussion of the main results achieved in this work is presented, along with the conclusions and some possible future work.

2 Methodology

The dynamic behavior of any kind of periodic structure, e.g., multi-span bridges and rotary machines, is commonly modeled exploiting the one-dimensional MDOF system formulation. Periodic structures identify those systems composed of the repetition of the same structural basic elements, i.e., spring, mass, and damper, eventually with different parameters, joint together. Moreover, particular elements may be introduced with the purpose of describing non-linear dynamics. In particular, the one-dimensional MDOF formulation is suitable for describing structures mainly experiencing motion in one single direction. Within this framework, damage is often simulated by reducing the stiffness value of one or more springs in the MDOF system [25, 36, 38]. In this work, it is assumed that acceleration measurements be available from sensors eventually installed on the real structure, thus allowing to estimate some hidden internal states of the structure, i.e., springs stiffnesses, by means of PF posterior estimation. Note, however, that the states estimation procedure would in principle work processing other type of measurements as well, e.g., masses displacements and velocities.

The differential equations of motion governing the dynamic behavior of the MDOF system degrees of freedom are implemented in the proposed procedure. Those are exploited for generating the state-space database describing the system observed behavior, thus allowing the PF-based state estimation process. It is assumed that the acceleration value of each degree of freedom of the system be measured at each time step in which the differential equations of motions are discretized. Then, the PF algorithm samples a large number of particles, i.e., plausible system dynamic evolutions, and further assigns to each possible trajectory an importance weight by a Bayesian combination with the observations on the structure under assessment. Finally, properly combining importance weights and particles, a posterior estimation of some hidden states, i.e., the MDOF model structural stiffnesses, is provided. In case no varying operating and environmental conditions affects the scenario, the PF framework by itself may be able to perform damage detection, localization, and quantification [27]. On the other hand, when confounding external agents are present, even the only and simple damage detection procedure by means of the PF estimation procedure would produce unsatisfactory results.

Hence, it is proposed in this work an original, fully integrated PF—auto-encoder framework, which leveraging the capability of the two combined methods allows to perform damage detection and localization even in presence of varying external conditions and non-linear system dynamics. A training phase of the auto-encoder with a healthy, temperature-dependent baseline allows the neural network to learn the unknown relationship between temperature and stiffness. Later on, in the so-called operational phase, damage diagnosis is achieved comparing the input data of the auto-encoder to the output data. Moreover, an automatic statistical threshold is introduced, with the purpose of establishing an automatic alarm system independent of any human interpretation of quantitative results.

PF algorithms and neural network auto-encoders are briefly introduced herein, focusing on their mutual interaction within the general framework established in this work. For further mathematical details on the individual methods, the interested reader is referred to the vast available literature.

2.1 *State-Space Model, Particle Filter an Auto-Encoders*

The proposed methodology refers to the model shown in Fig. 1, which represents a generic MDOF model of a mass-spring-damper system. Here the assumption is made that all the damping values be proportional to a stiffness reference value through a coefficient β, which is common to all the dampers in the system [40]. This assumption of proportional damping is

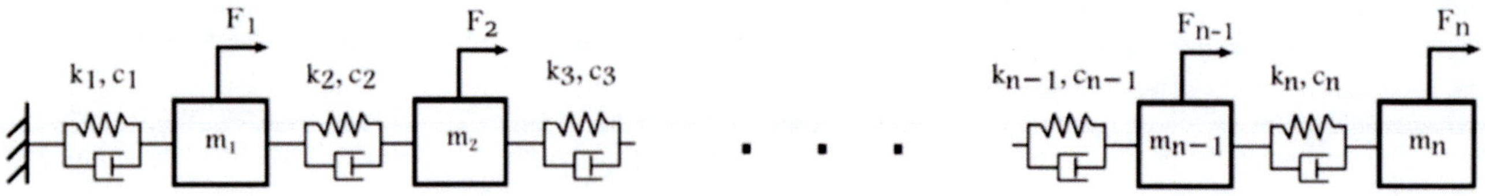

Fig. 1 MDOF mass-spring-damper system

commonly adopted in modeling structural systems [41]. Moreover, it is assumed that the first degree of freedom is affected by a degrading hysteretic behavior described by the Bouc–Wen's formula [28]:

$$\dot{r}(t) = \dot{x} - \beta_{BW}|\dot{x}||r|^{n-1}r - \gamma\dot{x}|r|^n \tag{1}$$

In order to setup the environment for the PF posterior estimation framework, an augmented state-space representation of the MDOF system considered is introduced. The position and the speed of each mass, the stiffness values, the unique proportionality damping constant, the Bouc-Wen parameters β_{BW}, γ and n, and the hysteretic displacement r are set as state variables. Hence, considering the generic case of n degrees of freedom:

$$\boldsymbol{x}^\tau = \left[x_1^\tau \dots x_n^\tau \; y_1^\tau \dots y_n^\tau \; k_1^\tau \dots k_n^\tau \; \beta \; \beta_{BW} \; \gamma \; n \; r\right]^\mathsf{T} \tag{2}$$

where x_i^τ is the position of mass i at time step τ, y_i^τ is the speed of mass i at the generic, discrete time step τ, k_i^τ is the i-th spring stiffness at time step τ, β is the proportionality constant of damping, according to the relationship $c_i^\tau = \beta k_i^\tau$, β_{BW}, γ and n are the Bouc-Wen parameters, and r the hysteretic displacement. The system evolution in time, evaluated at each discrete time step τ, is described by a state-space model which has to be a hidden Markov process of the generic form:

$$\boldsymbol{x}^\tau = g\left(\boldsymbol{x}^{\tau-1}, \boldsymbol{w}^{\tau-1}\right) \tag{3}$$

where $g(\cdot)$ represents a generic, possibly non-linear function and $\boldsymbol{w}^{\tau-1}$ is the process noise vector value at time step $\tau - 1$. The noise considered in this work is a Gaussian noise determined according to a multivariate distribution $N(0, \Sigma_w)$, $\Sigma_w = diag(\sigma_{x_1} \dots \sigma_\beta)$. In particular, considering the MDOF system affected by degrading hysteretic behavior adopted as reference system in the proposed framework, five state evolutions are identified, according to the state variables listed in (2). The position of each degree of freedom, i.e., mass, of the system evolves in time according to:

$$x_i^{\tau+1} = y_i^\tau dt + x_i^\tau + w_{position}^\tau; \quad \forall i = 1, \dots, n \tag{4}$$

That equation directly follows from the discretization in time of the system of differential equations of motion of the MDOF system, to which a position process noise $w_{position}^\tau$ is added. As a result of a similar procedure, the state-space equation governing the dynamic behavior of the velocity of each degree of freedom is:

$$\begin{aligned} y_i^{\tau+1} = {} & \frac{F_i^\tau}{m_i}dt + y_i^\tau(1 - \frac{dt}{m_i}(c_i^\tau + c_{i+1}^\tau)) + y_{i-1}^\tau\frac{dt}{m_i}c_i^\tau + y_{i+1}^\tau\frac{dt}{m_i}c_{i+1}^\tau - x_i^\tau\frac{dt}{m_i}(k_i^\tau + k_{i+1}^\tau) + \\ & + x_{i-1}^\tau\frac{dt}{m_i}k_i^\tau + x_{i+1}^\tau\frac{dt}{m_i}k_{i+1}^\tau + r_i^\tau k_i^\tau \delta_{i1} + w_{speed}^\tau; \quad \forall i = 1, \dots, n \end{aligned} \tag{5}$$

where F_i^τ represents a generic force acting on mass m_i, c_i is the damping coefficient value of the i-th damper, r_i is the hysteretic displacement pertaining to the i-th degree of freedom, and δ_{i1} is the Kronecker delta, which is introduced with the purpose of accounting for the fact that only the first degree of freedom is affected by degrading hysteretic behavior. Finally, w_{speed}^τ is the speed process noise. Equation (5) slightly changes for the first and the last degree of freedom of the system, since it must be updated canceling out the contribution of the missing components, i.e., the terms with x_{i-1}^τ and y_{i-1}^τ for $i = 1$, and those with x_{i+1}^τ, y_{i+1}^τ, k_{i+1}^τ and c_{i+1}^τ for $i = n$. The discretized dynamic equation governing the Bouc-Wen hysteretic displacement reads:

$$r^{\tau+1} = r^\tau + dt(y_1 - \beta_{BW}|y_1||r^\tau|^{n-1}r^\tau - \gamma(y_1)|r^\tau|^n) + w_r^\tau \tag{6}$$

where w_r^τ represents the hysteretic displacement process noise. Note that this equation is specialized to MDOF systems with degrading hysteretic behavior on the first degree of freedom only. All the variables which, in a healthy structure, are assumed to be invariant over time with respect to variations of external conditions other than temperature are modeled as an augmented state with constant value in time, to which some process noise is added to represent the effects of unmodeled dynamics. For instance, the equation governing the evolution of the spring stiffness value over time is:

$$k_i^{\tau+1} = k_i^\tau + w_{spring}^\tau; \quad \forall i = 1, \ldots, n+1 \tag{7}$$

where w_{spring}^τ is the process noise related to stiffness. Similar equations determine the evolution of the other invariant parameters included in the state-space, i.e., β, β_{BW}, γ, and n, with the only difference of replacing the process noise related to stiffness to the corresponding one, i.e., $w_{damping}^\tau$, $w_{\beta_{BW}}^\tau$, w_γ^τ, and w_n^τ. Moreover, stiffness is the only parameter which is assumed to vary with temperature. Hence, rescaling the relationship reported in [25] in order to comply with the stiffness values considered in this work, the discretized equation governing the stiffness evolution in time due to eventual change in temperature is represented by:

$$k_i(T) = (0.1\,(T)^2 - 18T + 10000) \cdot \frac{k_i(0°C)}{T}; \quad \forall i = 1, \ldots, n \tag{8}$$

where T identifies the external temperature measured in °C.

The equations shown above represent the evolution of the hidden Markov states of the MDOF system considered. Those hidden states are estimated within the PF framework processing the information from some measurements z, which are related to the system states through a model which is assumed to be known, which is determined by the following general function:

$$\boldsymbol{z}^\tau = h\left(\boldsymbol{x}^\tau, \boldsymbol{v}^\tau\right) \tag{9}$$

where $h(\cdot)$ is a generic, possibly non-linear function and $\boldsymbol{v}^\tau$ is the Gaussian measurement noise vector at time step τ, which is assumed to be distributed according to a multivariate distribution $N(0, \Sigma_v)$. Assuming that only acceleration measurements are available, the explicit form of Eq. (9) for the i-th degree of freedom at time step τ is:

$$\begin{aligned} z_i^\tau = & \frac{F_i^\tau}{m_i} - \frac{y_i^\tau}{m_i}(c_i^\tau + c_{i+1}^\tau) + \frac{y_{i-1}^\tau}{m_i} c_i^\tau + \frac{y_{i+1}^\tau}{m_i} c_{i+1}^\tau - \frac{x_i^\tau}{m_i}(k_i^\tau + k_{i+1}^\tau) + \frac{x_{i-1}^\tau}{m_i} k_i^\tau + \frac{x_{i+1}^\tau}{m_i} k_{i+1}^\tau + \\ & - \frac{r_i^\tau}{m_i} k_i^\tau \delta_{i1} + v_i^\tau; \quad \forall i = 1, \ldots, n \end{aligned} \tag{10}$$

where v_i^τ represents the acceleration measurement noise at time step τ, so that $\Sigma_v = diag(\sigma_{z_1} \ldots \sigma_{z_n})$. When the acceleration equation for the extreme degrees of freedom, i.e., $i = 1, n$, is considered, Eq. (10) has to be modified deleting those terms including x_{i-1}^τ and y_{i-1}^τ for $i = 1$, those with x_{i+1}^τ, y_{i+1}^τ, k_{i+1}^τ and c_{i+1}^τ for $i = n$.

The augmented state-space representation of the MDOF system presented above is considered within the PF posterior estimation framework for performing state estimation. Particle filtering represents a widely used model-based parameters identification and state estimation Bayesian method which can deal with complex scenarios, including non-Gaussian noise and non-linear dynamics. In particular, the selected PF algorithm is the sample importance resampling (SIR) algorithm, the main operative steps of which have already been described by the authors [42]. For brevity's sake, in this work no detailed mathematical treatment of the topic is given. The interested reader is referred to the works in [28, 43, 44] for further details about the functioning of the PF SIR algorithm.

However, as already mentioned above, exploiting PF posterior estimation only may not reveal successful in performing damage detection in case varying external and operational conditions arise, leading to inaccurate results. Thus, it is employed the original, unique framework already presented in the work in [42], to which the reader is referred to for the detailed description of the procedure adopted in this work. That framework consists of the integration of the PF posterior estimation with neural network auto-encoders, which allow to offset the effects of confounding factors from the damage detection and localization procedure. Auto-encoders are deep neural networks characterized by a symmetric structure of layers, consisting of one dimensionality reduction side, which reduces the input dimensionality, followed by a reconstructing side, which aims at producing in the output layer the same values fed in input. In particular, one auto-encoder per spring in the MDOF system is considered in this work, all sharing the same structure: one input layer, three hidden layers, i.e., mapping, bottleneck, and demapping layers, and one output layer (Fig. 2). The relationship between layers is set up during the so-called training phase,

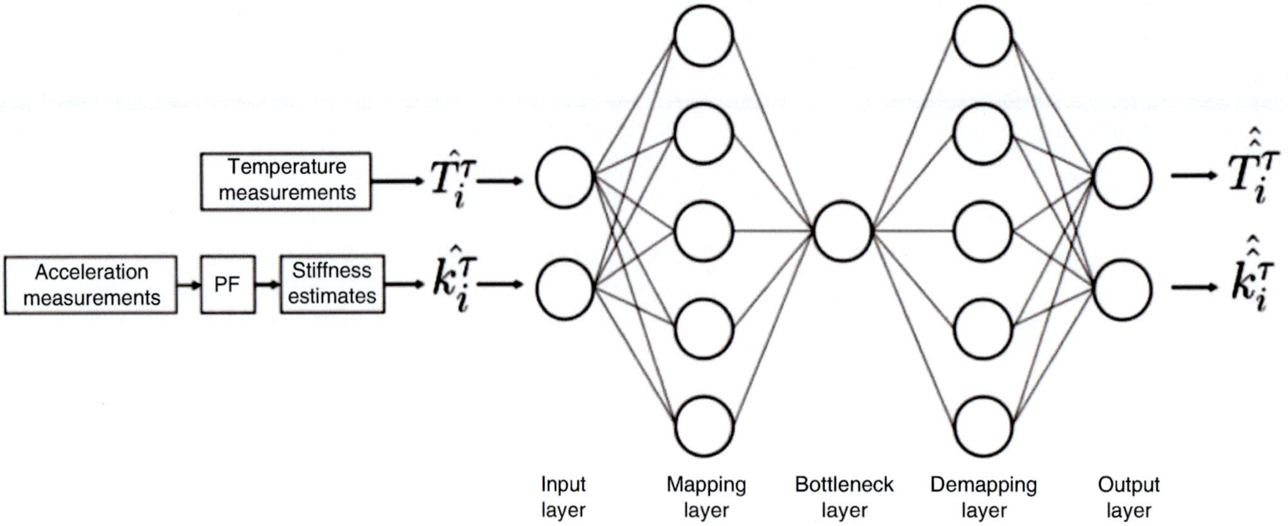

Fig. 2 Proposed auto-encoder structure

during which the implicit relationship between the inputs of the auto-encoder, i.e., temperature and stiffness, is learned by the network by analyzing supervised examples from a healthy baseline. During the operational phase of the damage assessment procedure, the auto-encoder receives as input the temperature value measured at the location of the respective spring, along with the PF-based estimate of the stiffness of that spring. The input signals are reconstructed in the output layer of the auto-encoder according to the model learned during the training phase. In case no damage occurs, the healthy relationship between temperature measurement and spring stiffness entirely drives the stiffness value change, allowing to perfectly reconstruct the input values in the output network layer. Differently, in case damage occurs, which is modeled as a temperature-uncorrelated stiffness value change, the auto-encoder may be likely to giving rise to some discrepancies between the input and the output data. The reconstruction error between output and input data is processed to detect eventual anomalies in the MDOF system that affects the healthy baseline correlation between temperature and stiffness values. Hence, as already mentioned above, for performing damage localization of the anomalies, in the general case of an n-degrees of freedom MDOF model, n auto-encoders are associated with the n springs in the system. Indeed, localization is allowed since only the reconstruction error of the auto-encoder associated with a damaged spring would be affected by the presence of the anomaly. According to previous literature works [36, 45, 46], the squared error between the inputs and the reconstructed outputs of the auto-encoders is defined as:

$$SE_i^\tau = (\hat{T}_i^\tau - \hat{\hat{T}}_i^\tau)^2 + (\hat{k}_i^\tau - \hat{\hat{k}}_i^\tau)^2 \tag{11}$$

where SE_i^τ is the error between the input variables (i.e., the temperature measurement $\hat{T}_i^\tau$ and the PF stiffness estimate $\hat{k}_i^\tau$) and the output variables ($\hat{\hat{T}}_i^\tau$ and $\hat{\hat{k}}_i^\tau$) recorded at time step τ on the *i-th* spring of the system. Neural networks need the data fed in input to be normalized in order to perform properly, in particular, this is vital when more inputs characterized by different units of measurement are analyzed. In this work, the normalization procedure of the input data is achieved by means of the following linear transformation:

$$\frac{\hat{T}_i^\tau - (-50\,^\circ\mathrm{C})}{120\,^\circ\mathrm{C} - (-50\,^\circ\mathrm{C})} \tag{12}$$

$$\frac{\hat{k}_i^\tau}{10\frac{N}{m}} \tag{13}$$

Hence, the squared error defined above can vary in the range $SE_i^\tau \in [0, 2]$. In what follows, the values $\hat{T}_i^\tau$ and $\hat{k}_i^\tau$ directly refer to the already normalized input data.

2.2 Definition of the Diagnostic Indicators

The squared error value related to each spring in the system has to be further processed in order to eventually trigger a diagnostic alarm. Those alarms may be given comparing the SE_i^τ value with a properly defined threshold, as suggested in [38]. However, the threshold value definition is arbitrary and may lead to inaccurate results. Moreover, in case of a not properly defined threshold, false alarms may be given since noise affects the inputs signals to the auto-encoder, due to a combination of acceleration measurement noise, process noises simulating unmodeled dynamics and Monte Carlo errors. Hence, in order to try to solve these issues, the authors proposed an original method consisting of combining an automatic threshold definition module with the construction of both a deterministic and a probabilistic indicator [42]. In particular, each spring $i = 1, \ldots, n+1$ is given one threshold value according to the following relationship:

$$ATSE_i^{99} = p99(SE_i^{1:\tau_0}) \tag{14}$$

where $p99(SE_i^{1:\tau_0})$ is the 99th percentile of the probability density function of the squared error defined in Eq. (11), computed over a fixed time window of width τ_0 time steps taken at the beginning of the operational life of the structure, when the system is assumed to be in healthy conditions.

An on/off deterministic fault indicator is introduced for each spring i, which is built up, at each time step i, comparing the $ATSE_{99}^i$ threshold value with the moving average over a sliding window of width τ_0 time steps of the squared errors for each spring i, i.e., $\mu_{SE_i^{\tau-\tau_0+1,\tau}}$. Note that the time window within which the automatic thresholds are defined coincides with the one over which the moving average is computed. Thanks to the properties of the moving average operation, the wider the window width, the more robust the algorithm to false alarms from error fluctuations, with the drawback of increasing the anomaly detection time. The deterministic fault indicator is defined according to the following relationship:

$$Id_{SE,i}^{\tau-\tau_0+1,\tau} = \begin{cases} 1 & \text{if } \mu_{SE_i^{\tau-\tau_0+1,\tau}} > ATSE_i^{99} \\ 0 & \text{otherwise} \end{cases} \tag{15}$$

Hence, the fault indicator is activated, i.e., is set to 1, whenever the signal $\mu_{SE_i^{\tau-\tau_0+1,\tau}}$ exits the healthy region defined by the respective threshold. Therefore, only significant variations of the signal $\mu_{SE_i^{\tau-\tau_0+1,\tau}}$ may trigger alarms, filtering out the confounding effects coming from noise and disturbances.

Moreover, an additional, probabilistic fault indicator is introduced in order to further damp out possible false alarms triggered by its deterministic counterpart:

$$Ip_{SE,i}^{\tau-\tau_0+1,\tau} = \frac{\sum_{j=\tau-\tau_0+1}^{\tau} Id_{SE,i}^{j}}{\tau_0} \tag{16}$$

where $Id_{SE,i}^\tau$ represents the instantaneous counterpart at time step τ of $Id_{SE,i}^{\tau-\tau_0+1,\tau}$, which is defined as:

$$Id_{SE,i}^{\tau} = \begin{cases} 1 & \text{if } SE_i^\tau > ATSE_i^{99} \\ 0 & \text{otherwise} \end{cases} \tag{17}$$

3 Case Study

The case study presented in this work is developed considering a three degrees of freedom MDOF system on the basis of the one shown in Fig. 1. The system parameters are taken from the work in [28], according to which $m_i = 1kg\ \forall i = 1, \ldots, 3$, $k_j = 9\dfrac{N}{m}$ and $c_j = 0.25\dfrac{N \cdot s}{m}$ (i.e., $\beta = 27.8 \cdot 10^{-3}$) $\forall j = 1, \ldots, 3$). The first degree of freedom is assumed to be affected by non-linear hysteretic degradation with parameters $\beta_{BW} = 2$, $\gamma = 1$ and $n = 2$. The structural state variables are affected by normally distributed process noise characterized by zero mean and variances $\sigma^2_{w,position} = 10^{-16}\,\mathrm{m}^2$, $\sigma^2_{w,speed} = 10^{-16}\frac{\mathrm{m}^2}{\mathrm{s}^2}$, $\sigma^2_{w,spring} = 1.62 \cdot 10^{-5}\frac{\mathrm{N}^2}{\mathrm{m}^2}$, and $\sigma^2_{w,damping} = 8 \cdot 10^{-13}$ for position, velocity, stiffness, and damping proportionality constant, respectively. The process noise related to the Bouc-Wen model adimensional parameters is characterized by variances $\sigma^2_{w,\beta_{BW}} = 9.6 \cdot 10^{-7}$, $\sigma^2_{w,\gamma} = 2.4 \cdot 10^{-7}$ and $\sigma^2_{w,n} = 9.6 \cdot 10^{-7}$, while the hysteretic displacement noise is assigned the variance $\sigma^2_{w,r} = 10^{-16}\,\mathrm{m}^2$. Similarly, the observation noise related to the only measured variable considered in this work, i.e., acceleration, is taken as $\sigma^2_v = 1 \cdot 10^{-5}\,\frac{\mathrm{m}^2}{\mathrm{s}^4}$. The widely adopted practice of simulating the response to earthquakes using white noise accelerations to generate the forces acting on the system degrees of freedom is applied in this case study [33]. Northridge earthquake-like forces are considered as forcing terms, distributed as white noise with maximum excitation frequency of 30 Hz [28], absolute value acceleration amplitude in the range $[0g; 1.8g]$ [47] and time duration equal to the analysis time. The analysis time horizon is $T_{hor} = 16.7$ s, discretized in $N_t = 3 \cdot 10^4$ time intervals of width $dt = 0.556$ ms each. The assumption is made that the structural elements are in equilibrium with the external temperature, which linearly decreases from the initial value 40 °C to the final one 10 °C within the analysis time horizon T_{hor}. Damage is introduced in spring number 3 as a progressive linear degradation of the corresponding stiffness value, occurring at time $t = 8.3$, up to a final stiffness value at the analysis time horizon equal to the 70% of the corresponding healthy one at the same temperature, i.e., at 10 °C. The sampling frequency is $f_s = 1/dt = 1800$ Hz, which turns out to satisfy the Shannon sampling theorem, being sufficiently larger than the forcing function frequency $f_{force} = 30$ Hz.

A PF with $3 \cdot 10^4$ particles is used for estimating the whole state vector $\boldsymbol{x}^\tau$ at each time step τ. The particles' values are initialized as follows:

- the position and velocity states estimates are initialized according to a normal distribution with zero mean and variance $\sigma^2_{w,position} = 10^{-16}\,\mathrm{m}^2$ and $\sigma^2_{w,speed} = 10^{-16}\,\frac{m^2}{s^2}$, respectively;
- the stiffnesses, the damping coefficient, the Bouc-Wen model parameters, and the hysteretic displacement are initialized to a random number between the 95% and the 105% of the actual state value to be estimated.

The auto-encoders employed in the simulation need to be trained with a healthy baseline acquired on the system under assessment. That baseline is built up simulating the non-damaged system and estimating the model state-space with the PF. A time horizon $T^{baseline}_{hor} = 20$ s is considered for the baseline acquisition, simulated by means of $N^{baseline}_t = 3 \cdot 10^4$ time steps of width $dt^{baseline} = 0.667$ ms each. The structural elements are assumed to be in instantaneous equilibrium with the external temperature value, common to all the elements in the system, linearly increasing from −50 to 120 °C. This wide temperature range is selected for making sure that in the operational phase of the SHM framework the auto-encoder is fed with temperature values inside the range it has been trained with. Indeed, this is only possible when a real temperature-controlled environment is considered or in artificially generated case studies. In real applications with no temperature control available, a long baseline needs to be acquired to ensure the possible temperature range is sufficiently represented. Since all the springs are identical and experience the same temperature value at the same time instant, i.e., no temperature spatial gradient is considered, it turns out it is enough to train one auto-encoder considering any spring in the MDOF system. Note that, in the operational phase, one copy of the trained auto-encoder per spring is considered, in order to allow damage localization. The temperature observations are generated at time step τ using $\hat{T}^\tau = T^\tau + w^\tau_T$, where $w^\tau_T \sim N(0, 0.1)$°C is the corresponding observation noise considered herein. Thus, a total of $N^{baseline}_t$ examples are generated and used for training the auto-encoder exploited in this work, the layout of which is described in Sect. 2.1. The MATLAB Deep Learning Toolbox™ is employed for building and training the auto-encoder. After the training phase it is possible to put on-line one auto-encoder per spring and to start with the operational phase.

The state-space estimates of the spring stiffnesses are shown in Fig. 3, where the estimates are identified with red lines and the corresponding reference values with black ones. As already mentioned above, the estimated value of each stiffness is not initialized to the actual one. It can be seen that the PF estimate goes to the reference value with quite fast settling time, with satisfactory parameters identification performance. The same consideration also holds for the stiffness degradation due to damage, which is clearly identified. Note that in the present case study it is possible to identify the damaged spring just by looking at the PF estimate value (Fig. 3c). The framework further processes the PF-estimated stiffness values and the temperature measurements, corrupted with noise according to the healthy baseline construction procedure outlined above, which are then fed in real time to the three auto-encoders.

The squared error SE^τ_i between the input and output values of the auto-encoder (solid lines), evaluated by means of Eq. (11), its average over the time window of width $\tau_0 = 5000$ time steps, corresponding to a time horizon of $2.8s$, i.e., $\mu_{SE^{\tau-\tau_0+1,\tau}_i}$ (black, dashed lines), and the corresponding automatic threshold $ATSE^{99}_i$ (red dotted line), for the stiffnesses k_1, k_2, and k_3 are shown in Fig. 4. The thresholds are only shown after the first τ_0 time steps, which are required to automatically identify their values, under the assumption that the system is healthy. Figure 5 presents the deterministic fault indicator time history for each spring in the system, i.e., $Id^{\tau-\tau_0+1,\tau}_{SE,i}$, while the probabilistic counterparts $Ip^{\tau-\tau_0+1,\tau}_{SE,i}$ are shown in Fig. 6. Spring $i = 1$ is well characterized by the proposed framework. According to the fact that no damage is introduced on that

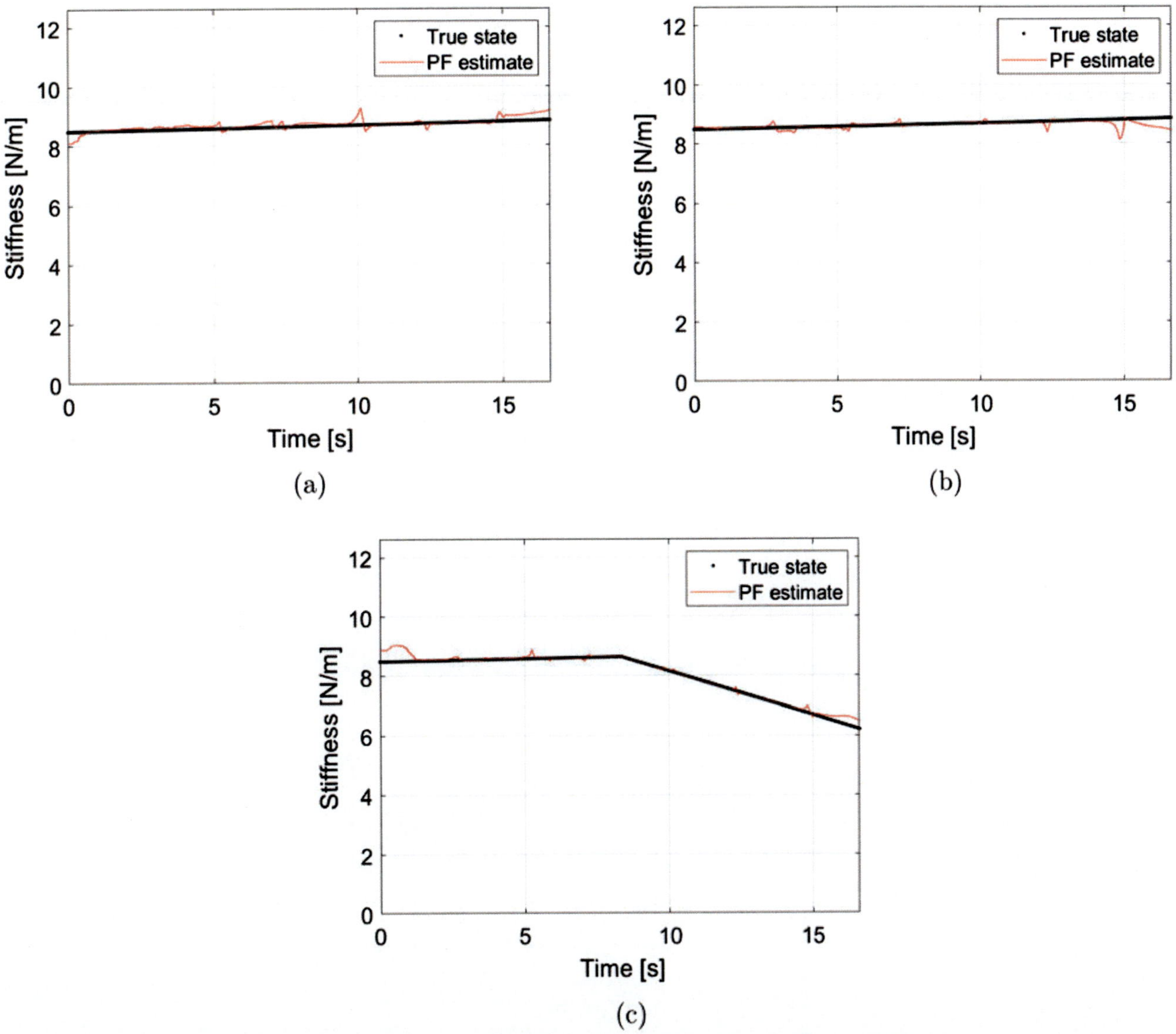

Fig. 3 Temporal evolution of the particle filter stiffness estimates (red) and true stiffness values (black) of spring $i = 1$ (**a**), $i = 2$ (**b**), $i = 3$ (**c**)

spring, the mean squared error SE_1^τ stays within the healthy region defined by the threshold $ATSE_1^{99}$ (Fig. 4a). Accordingly, the deterministic index $Id_{SE,1}^{\tau-\tau_0+1,\tau}$ is never triggered (Fig. 5a). The probability of damage occurrence $Ip_{SE,1}^{\tau-\tau_0+1,\tau}$ is always equal to zero, except from a slight increase from time $t = 10.1$ s to $t = 12.9$ s, determined by the fact that there is a peak instantaneous value of the squared error related to spring $i = 1$ that temporarily exits the healthy region at $t = 10.1$ s. A similar behavior is identified in the other healthy spring, i.e., spring $i = 2$. However, in the last second of analysis the mean squared error SE_2^τ exits the corresponding healthy region (Fig. 4b), determining a false alarm in the deterministic fault indicator $Id_{SE,2}^{\tau-\tau_0+1,\tau}$. Due to two wide oscillations of the instantaneous squared error value associated with spring $i = 2$, which occur at time instants $t = 12.3$ s and $t = 14.7$ s, the probabilistic damage indicator $Ip_{SE,2}^{\tau-\tau_0+1,\tau}$ starts identifying a possible failure at $t = 12.3$ s. However, no sensible failure probability is reached, being the maximum registered value equal to 39%. Similarly, also spring $i = 3$ is correctly identified by the proposed methodology. The progressive degradation damage in introduced in the model at time $t = 8.3$ s and identified at $t = 11.5$ s. In fact, at that time instant the mean squared error value SE_3^τ overcomes the safety threshold $ATSE_3^{99}$ triggering the deterministic fault indicator $Id_{SE,3}^{\tau-\tau_0+1,\tau}$, which is kept on the alert level until the end of the analysis. Accordingly, the probabilistic damage indicator $Ip_{SE,2}^{\tau-\tau_0+1,\tau}$ reaches the peak value, i.e., 100% failure probability, at time $t = 13.1$ s, without decreasing as time goes by.

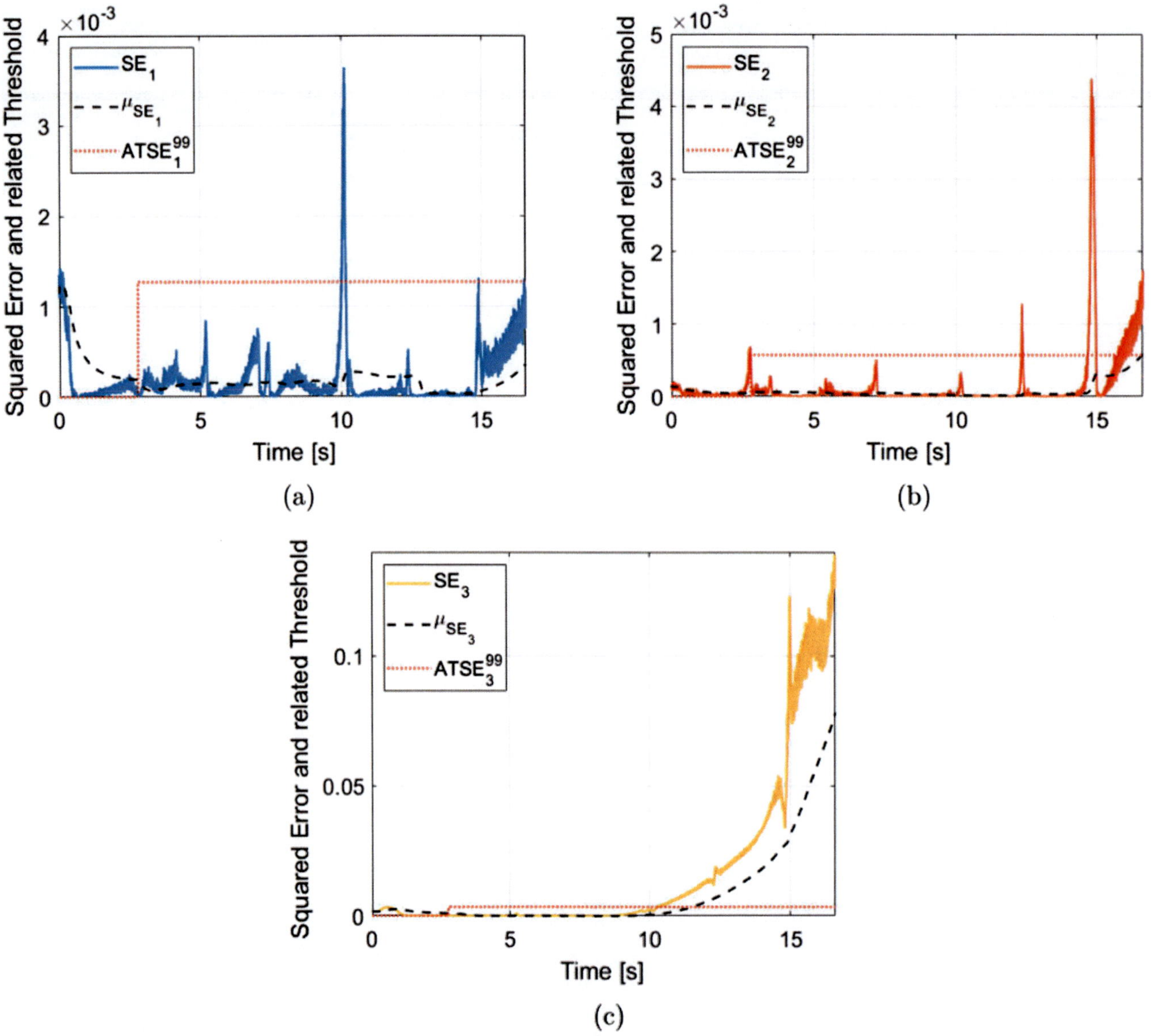

Fig. 4 Temporal evolution of the squared error SE_i^τ between input and output layers of the auto-encoder (solid lines) for k_1, k_2 and k_3, the colored dashed lines indicate the corresponding automatic threshold. (**a**) Spring 1. (**b**) Spring 2. (**c**) Spring 3

4 Conclusions

An original framework combining PF and auto-encoders is proposed in this study, with the aim of providing a tool suitable for performing damage diagnosis in structures subject to changing environmental conditions. The framework is specialized to the general case of mono-dimensional MDOF systems characterized by non-linear hysteretic behavior, and it is applied in Sect. 3 to a 3DOF system, proving that it satisfactorily provides both real-time estimation of each spring stiffness and information about the presence of eventual damages, along with their position.

Eventual changes in the stiffness value of a spring in the system are provided by the particle filter embedded in the framework; however, it is not able to provide any insight in the cause of those variations, which may be caused either by damages or by changing external conditions. Hence, one auto-encoder per spring receives the stiffness estimate from the PF and offsets eventual temperature variation effects from it. In order to do so, the auto-encoder needs to be trained with a healthy baseline properly acquired to describe well the range of operating conditions the system is supposed to work in. In the operational phase, the squared error mean value between the input and the output of each auto-encoder is employed to detect any anomaly which may arise from eventual damages. The squared error mean value computed over a properly set

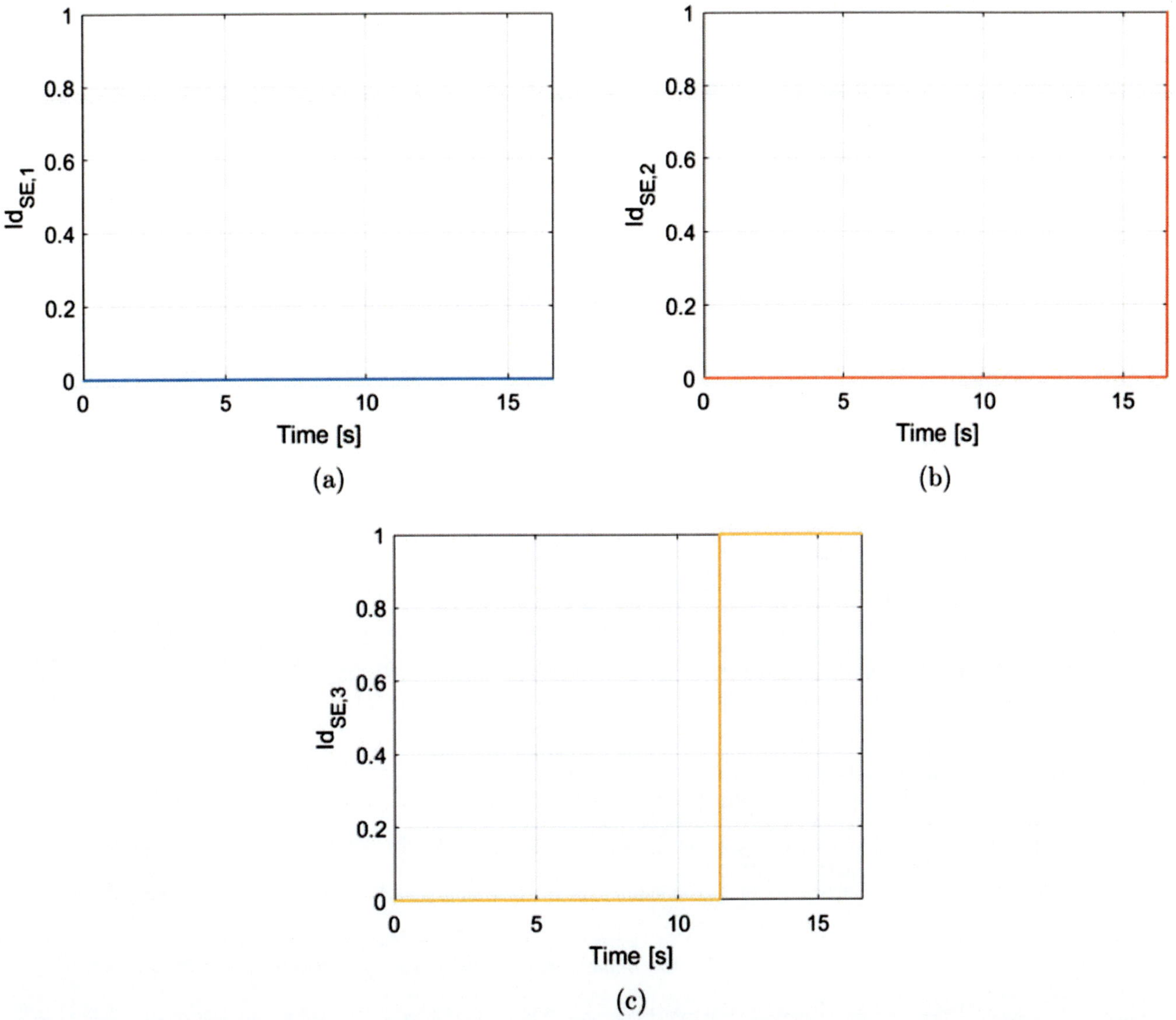

Fig. 5 Temporal evolution of fault indicator $Id_{SE,i}^{\tau-\tau_0+1,\tau}$ for springs 1 (**a**), 2 (**b**) and 3 (**c**)

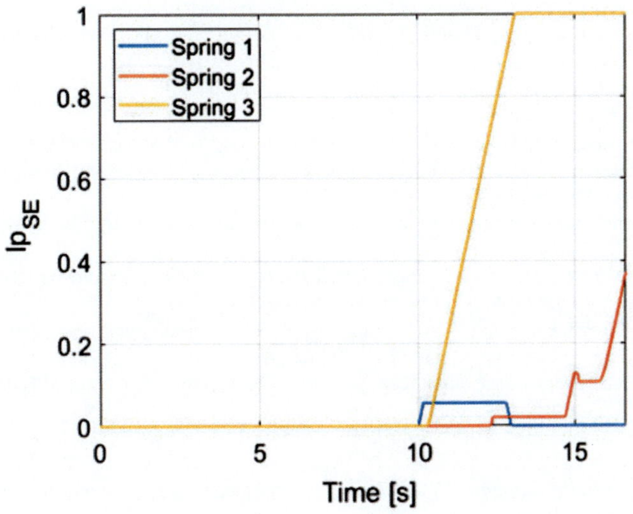

Fig. 6 Temporal evolution of the probabilistic fault indicator $Ip_{SE,i}^{\tau-\tau_0+1,\tau}$

moving time window is compared to an automatically set threshold for properly turning on the fault indicator $Id_{SE,i}^{\tau-\tau_0+1,\tau}$. Moreover, in order to damp out the false alarms due to eventual oscillations of the PF estimate, since increasing the particles involved in the simulation would significantly increase the required computational resources, a more robust probabilistic fault indicator $Ip_{SE,i}^{\tau-\tau_0+1,\tau}$ is also computed for each spring i in the MDOF system, with overall satisfactory performances. As shown in the case study reported in Sect. 3, the deterministic fault indicator occasionally gives false alarms, while the probabilistic indicators provide more reliable information about eventual damages in the structure assessed.

Further work still has to be done in this field. Expanding the proposed framework for considering changing external conditions different than temperature would provide interesting, more general results. Moreover, the algorithm may be further enhanced in order to perform damage diagnosis using as input values only measured system output responses, i.e., basing the diagnosis task only on vibration signals induced by environmental/operative causes (wind, traffic, etc.). This would allow avoiding the need for artificial forcing functions, which might turn out to be complex and expensive to accurately realize when dealing with real structures. To conclude, the effectiveness of the proposed methodology with non-Gaussian noise could also be investigated.

References

1. Frangopol, D.M., Liu, M.: Maintenance and management of civil infrastructure based on condition, safety, optimization, and life-cycle cost. Struct. Infrastruct. Eng. **3**(1), 29–41 (2007). https://doi.org/10.1080/15732470500253164
2. Artigao, E., Martin-Martinez, S., Honrubia-Escribano, A., Gomez-Lazaro, E.: Wind turbine reliability: a comprehensive review towards effective condition monitoring development. Appl. Energy **228**, 1569–1583 (2018). https://doi.org/10.1016/j.apenergy.2018.07.037. http://www.sciencedirect.com/science/article/pii/S0306261918310651
3. Ziegler, L., Gonzalez, E., Rubert, T., Smolka, U., Melero, J.J.: Lifetime extension of onshore wind turbines: a review covering germany, spain, denmark, and the uk. Renew. Sust. Energ. Rev. **82**, 1261–1271 (2018). https://doi.org/10.1016/j.rser.2017.09.100. http://www.sciencedirect.com/science/article/pii/S1364032117313503
4. Courrech, J., Gaudet, M.: Envelope analysis - the key to rolling-element bearing diagnosis. Bruel & Kjaer internal document Available online at https://www.bksv.com/media/doc/BO0187.pdf
5. Dackermann, U., Smith, W.A., Randall, R.B.: Damage identification based on response-only measurements using cepstrum analysis and artificial neural networks. Struct. Health Monit. **13**(4), 430–444 (2014)
6. Mechefske, C.: Machine condition monitoring and fault diagnostics. In: C.W. de Silva (ed.) Vibration and Shock Handbook, chap. 25. CRC Press, Taylor and Francis Group, Boca Raton, FL (2005)
7. Johnson, T.J., Adams, D.E.: Transmissibility as a differential indicator of structural damage. J. Vib. Acoust. **124** (2002). https://doi.org/10.1115/1.1500744
8. Maia, N.M., Almeida, R.A., Urgueira, A.P., Sampaio, R.P.: Damage detection and quantification using transmissibility. Mech. Syst. Signal Process. **25**(7), 2475–2483 (2011). https://doi.org/10.1016/j.ymssp.2011.04.002. http://www.sciencedirect.com/science/article/pii/S0888327011001567
9. Manson, G., Worden, K.: Experimental validation of a structural health monitoring methodology: Part iii. Damage location on an aircraft wing. J. Sound Vib. **259**(2), 365–385 (2003)
10. Chesné, S., Deraemaeker, A.: Damage localization using transmissibility functions: a critical review. Mech. Syst. Signal Process. **38**, 569–584 (2013). https://doi.org/10.1016/j.ymssp.2013.01.020. https://www.sciencedirect.com/science/article/pii/S0888327013000745
11. Sohn, H.: Effects of environmental and operational variability on structural health monitoring. Philos. Trans. R. Soc. Lond. A Math. Phys. Eng. Sci. **365**(1851), 539–560 (2007)
12. Farrar, C., Baker, W., Bell, T., Cone, K., Darling, T., Duffey, T., Eklund, A., Migliori, A.: Dynamic characterization and damage detection in the i-40 bridge over the rio grande. Technical Report (1994). https://doi.org/10.2172/10158042
13. Farrar, C.R., Doebling, S.W., Cornwell, P.J., Straser, E.G.: Variability of modal parameters measured on the Alamosa Canyon Bridge. In: Proceedings of International Modal Analysis Conference, vol. 1 (1997)
14. Ruotolo, R., Surace, C.: Damage detection using singular value decomposition. In: Proceedings of DAMAS, vol. 97 (1997)
15. Yan, A.M., Kerschen, G., De Boe, P., Golinval, J.C.: Structural damage diagnosis under varying environmental conditions - Part II: local PCA for non-linear cases. Mech. Syst. Signal Process. **19**(4), 865–880 (2005)
16. Kullaa, J.: Is temperature measurement essential in structural health monitoring? In: Proc. 4th Int. Workshop on Structural Health Monitoring, pp. 717–724 (2003)
17. Cross, E.J., Worden, K., Chen, Q.: Cointegration: a novel approach for the removal of environmental trends in structural health monitoring data. Proc. R. Soc. Lond. A Math. Phys. Eng. Sci. **467**(2133), 2712–2732 (2011)
18. Shi, H., Worden, K., Cross, E.J.: A regime-switching cointegration approach for removing environmental and operational variations in structural health monitoring. Mech. Syst. Signal Process. **103**, 381–397 (2018)
19. Sohn, H., Worden, K., Farrar, C.R.: Statistical damage classification under changing environmental and operational conditions. J. Intell. Mater. Syst. Struct. **13**(9), 561–574 (2002)
20. Peeters, B., De Roeck, G.: One-year monitoring of the z24-bridge: environmental effects versus damage events. Earthq. Eng. Struct. Dyn. **30**(2), 149–171 (2001)
21. Worden, K., Sohn, H., Farrar, C.R.: Novelty detection in a changing environment: regression and interpolation approaches. J. Sound Vib. **258**(4), 741–761 (2002)

22. Fritzen, C.P., Mengelkamp, G., Guemes, A.: Elimination of temperature effects on damage detection within a smart structure concept. Struct. Heal. Monit. **10**, 15–17 (2003)
23. Worden, K., Cross, E.J.: On switching response surface models, with applications to the structural health monitoring of bridges. Mech. Syst. Signal Process. **98**, 139–156 (2018)
24. Limongelli, M.: Frequency response function interpolation for damage detection under changing environment. Mech. Syst. Signal Process. **24**(8), 2898–2913 (2010). https://doi.org/10.1016/j.ymssp.2010.03.004. http://www.sciencedirect.com/science/article/pii/S0888327010000658
25. Soo Lon Wah, W., Chen, Y.-T.: A new approach toward damage localization and quantification of structures under changing temperature condition. J. Low Freq. Noise Vib. Active Control (2018). https://doi.org/10.1177/1461348418793079
26. Abdessalem, A.B., Dervilis, N., Wagg, D., Worden, K.: Model selection and parameter estimation in structural dynamics using approximate Bayesian computation. Mech. Syst. Signal Process. **99**, 306–325 (2018). https://doi.org/10.1016/j.ymssp.2017.06.017. http://www.sciencedirect.com/science/article/pii/S0888327017303321
27. Wan, C., Huang, R., Huang, L., Wen, B., Sato, T.: Damage identification using particle filters. Procedia Eng. **188**, 41–47 (2017). https://doi.org/10.1016/j.proeng.2017.04.455. http://www.sciencedirect.com/science/article/pii/S1877705817320076. Structural Health Monitoring - From Sensing to Diagnosis and Prognosis
28. Chatzi, E., Smyth, A.: The unscented Kalman filter and particle filter methods for nonlinear structural system identification with non-collocated heterogeneous sensing. Struct. Control. Health Monit. **16**, 99–123 (2009). https://doi.org/10.1002/stc.290
29. Kwak, H.-G., Kim, S.-P., Kim, J.-E.: Nonlinear dynamic analysis of RC frames using cyclic moment curvature relation. Struct. Eng. Mech. **17**(3-4), 357–378 (2004)
30. Loh, C.-H., Mao, C.H., Chao, S.-H., Weng, J.-H.: Feature extraction and system identification of reinforced concrete structures considering degrading hysteresis. Struct. Control. Health Monit. **17**(7), 712–729 (2010)
31. Loh, C.-H., Mao, C.H., Huang, J.-R., Pan, T.-C.: System identification of degrading hysteresis of reinforced concrete frames. Earthq. Eng. Struct. Dyn. **40**(6), 623–640 (2011)
32. Baber, T., Wen, Y.: Random vibration hysteretic, degrading systems. J .Eng. Mech. Div. **107**(6), 1069–1087 (1981)
33. Yang, J.N., Xia, Y., Loh, C.-H.: Damage detection of hysteretic structures with a pinching effect. J. Eng. Mech. **140**(3), 462–472 (2014)
34. Pathirage, C.S.N., Li, J., Li, L., Hao, H., Liu, W., Ni, P.: Structural damage identification based on autoencoder neural networks and deep learning. Eng. Struct. **172**, 13–28 (2018). https://doi.org/10.1016/j.engstruct.2018.05.109. http://www.sciencedirect.com/science/article/pii/S0141029618302062
35. Yam, L., Yan, Y., Jiang, J.: Vibration-based damage detection for composite structures using wavelet transform and neural network identification. Compos. Struct. **60**(4), 403–412 (2003). https://doi.org/10.1016/S0263-8223(03)00023-0. http://www.sciencedirect.com/science/article/pii/S0263822303000230
36. Sohn, H., Worden, K., Farrar, C.R.: Novelty detection using auto-associative neural network. In: ASME International Mechanical Engineering Congress and Exposition, Proceedings, vol. 2, pp. 1525–1532 (2001). www.scopus.com
37. Chang, C.M., Lin, T.K., Chang, C.W.: Applications of neural network models for structural health monitoring based on derived modal properties. Measurement **129**, 457–470 (2018). https://doi.org/10.1016/j.measurement.2018.07.051. http://www.sciencedirect.com/science/article/pii/S0263224118306559
38. Worden, K.: Structural fault detection using a novelty measure. J. Sound Vib. **201**(1), 85–101 (1997). https://doi.org/10.1006/jsvi.1996.0747. http://www.sciencedirect.com/science/article/pii/S0022460X96907476
39. Chesne, S., Deraemaeker, A.: Damage localization using transmissibility functions: a critical review. Mech. Syst. Signal Process. **38**(2), 569–584 (2013). https://doi.org/10.1016/j.ymssp.2013.01.020. http://www.sciencedirect.com/science/article/pii/S0888327013000745
40. Cheng, L., Cigada, A.: An analytical perspective about structural damage identification based on transmissibility function. Struct. Health Monit. **19**(1), 142–155 (2020). https://doi.org/10.1177/1475921719838079
41. Rayleigh, J.W.S.B.: The Theory of Sound. Macmillan, New York (1896)
42. Cadini, F., Lomazzi, L., Ferrater Roca, M., Sbarufatti, C., Giglio, M.: Neutralization of temperature effects in damage diagnosis of MDOF systems by combinations of autoencoders and particle filters. Mech. Syst. Signal Process. **162**, 108048 (2022). https://doi.org/10.1016/j.ymssp.2021.108048
43. Tulsyan, A., Gopaluni, B., Khare, S.: Particle filtering without tears: a primer for beginners. Comput. Chem. Eng. **95**, 130–145 (2016). https://doi.org/10.1016/j.compchemeng.2016.08.015. http://www.sciencedirect.com/science/article/pii/S0098135416302769
44. Arulampalam, M.S., Maskell, S., Gordon, N., Clapp, T.: A tutorial on particle filters for online nonlinear/non-gaussian bayesian tracking. IEEE Trans. Signal Process. **50**(2), 174–188 (2002). https://doi.org/10.1109/78.978374
45. Kramer, M.A.: Nonlinear principal component analysis using autoassociative neural networks. AIChE J. **37**(2), 233–243 (1991)
46. Harkat, M.F., Djelel, S., Doghmane, N., Benouaret, M.: Sensor fault detection, isolation and reconstruction using nonlinear principal component analysis. Int. J. Autom. Comput. **4**(2), 149–155 (2007)
47. Luco, N.: Seismic design and hazard maps: before and after. In: Structure Magazine, pp. 28–30 (2019)

A New Benchmark Problem for Structural Damage Detection: Bolt Loosening Tests on a Large-Scale Laboratory Structure

Onur Avci, Osama Abdeljaber, Serkan Kiranyaz, Mohammed Hussein, Moncef Gabbouj, and Daniel Inman

Abstract Monitoring the structural performance of engineering structures has always been pertinent for maintaining structural health and assessing the life cycle of structures. Structural Health Monitoring (SHM) and Structural Damage Detection (SDD) fields have been topics of ongoing research over the years to explore and verify different monitoring techniques and damage detection and localization procedures. In an attempt to compare performances of different methods, benchmark datasets are valuable resources since the data is made available to researchers enabling side-by-side comparisons. This paper presents a new experimental benchmark dataset generated from tests on a large-scale laboratory structure. The primary goal of the authors was to explore brand-new damage detection and quantification methodologies for efficient monitoring of structures. For this purpose, a large-scale steel grid structure with footprint dimensions of 4.2 m × 4.2 m was constructed in laboratory environment and it has been used as a test bed by the authors. The structural members of the structure are all IPE120 hot-rolled steel cross sections. The simulation of structural damage was simply loosening the bolts at one of the beam-to-girder connections, which is a slight change of rotational stiffness at the joint of the steel grid structure. The authors shared the dataset for 1 undamaged and 30 damaged conditions and published it on a public website as a new benchmark problem for structural damage detection at http://www.structuralvibration.com/benchmark/ so that other researchers can use the data and test algorithms. The authors also shared one of the damage detection tools they used, One-Dimensional Convolutional Neural Networks (1D-CNNs). The application codes, configuration files, and accompanied components of the 1D-CNNs package are available for viewers at http://www.structuralvibration.com/cnns/.

Keywords Benchmark problem · Structural Damage Detection · Structural Health Monitoring · Damage localization · Damage quantification · Damage identification · Steel structures

O. Avci (✉)
Civil, Construction and Environmental Engineering, Iowa State University, Ames, IA, USA
e-mail: oavci@vt.edu; oavci@iastate.edu

O. Abdeljaber
Department of Building Technology, Linnaeus University, Växjö, Sweden
e-mail: osama.abdeljaber@lnu.se

S. Kiranyaz
Department of Electrical Engineering, Qatar University, Doha, Qatar
e-mail: mkiranyaz@qu.edu.qa

M. Hussein
Department of Civil and Architectural Engineering, Qatar University, Doha, Qatar
e-mail: mhussein@qu.edu.qa

M. Gabbouj
Department of Computing Sciences, Tampere University, Tampere, Finland
e-mail: moncef.gabbouj@tuni.fi

D. Inman
Department of Aerospace Engineering, University of Michigan, Ann Arbor, MI, USA
e-mail: daninman@umich.edu

K. Grimmelsman (ed.), *Dynamics of Civil Structures, Volume 2*, Conference Proceedings of the Society for Experimental Mechanics Series, https://doi.org/10.1007/978-3-030-77143-0_2

1 Introduction

The traditional way of monitoring civil infrastructure is based on visual inspections [1, 2]. Along with the computing and sensing technologies, dynamic response of structures has started to be used for condition evaluation [3–11] and serviceability assessment [12–20] of engineering structures. Structural Health Monitoring (SHM) and Structural Damage Detection (SDD) have emerged as research fields targeting efficient monitoring of structures [21–29] for which Machine Learning (ML) and Deep Learning (DL) algorithms have been implemented [30–38] to facilitate and optimize the identification and localization processes. Before the researchers apply newly developed damage detection and damage quantification methods on real structures, it is important that they verify the methodologies experimentally in laboratory. Yet, building a laboratory structure, arranging the logistics for data collection and processing, is not always easy. This is why benchmark datasets are made available to researchers to benefit from existing data. Benchmark datasets are useful resources for comparing the performances of various damage detection/localization/quantification procedures. Such datasets are convenient because the data becomes accessible to all, and since everybody uses the same data, the performance comparisons are systematic and more efficient.

In this paper, the authors are presenting a new experimental benchmark dataset generated by tests on a large-scale laboratory structure. The primary goal of the authors was to explore and test brand-new damage detection and quantification methodologies on the laboratory structure. It was later decided to share the dataset in a benchmark format so that other researchers can use the data and test their own methodologies. The One-Dimensional Convolutional Neural Networks (1D-CNNs) code package was also published on a website about the same time, by the authors.

The benchmark structure used to simulate the damage conditions was an inclined steel grid structure with footprint dimensions of 4.2 m × 4.2 m (Fig. 1). By loosening the bolts at the beam-to-girder connections, 31 different damage conditions were created on the benchmark structure. Acceleration data was carefully collected for each damaged and undamaged condition. The collection of the acceleration recordings formed the Benchmark dataset, providing SHM and SDD researchers with a new test bed for verifying newly developed vibration-based damage detection algorithms.

The steel frame consists of hot-rolled members (4 columns, 25 filler beams, 8 girders and brace elements to stabilize the columns). While the length of the 20 typical filler beams is 77 cm, the 5 cantilevered filler beams are about 1 m long (Fig. 1). The length of the girders is 4.6 m. As per grandstand specifications, there is a 20° inclination on the structure. All structural

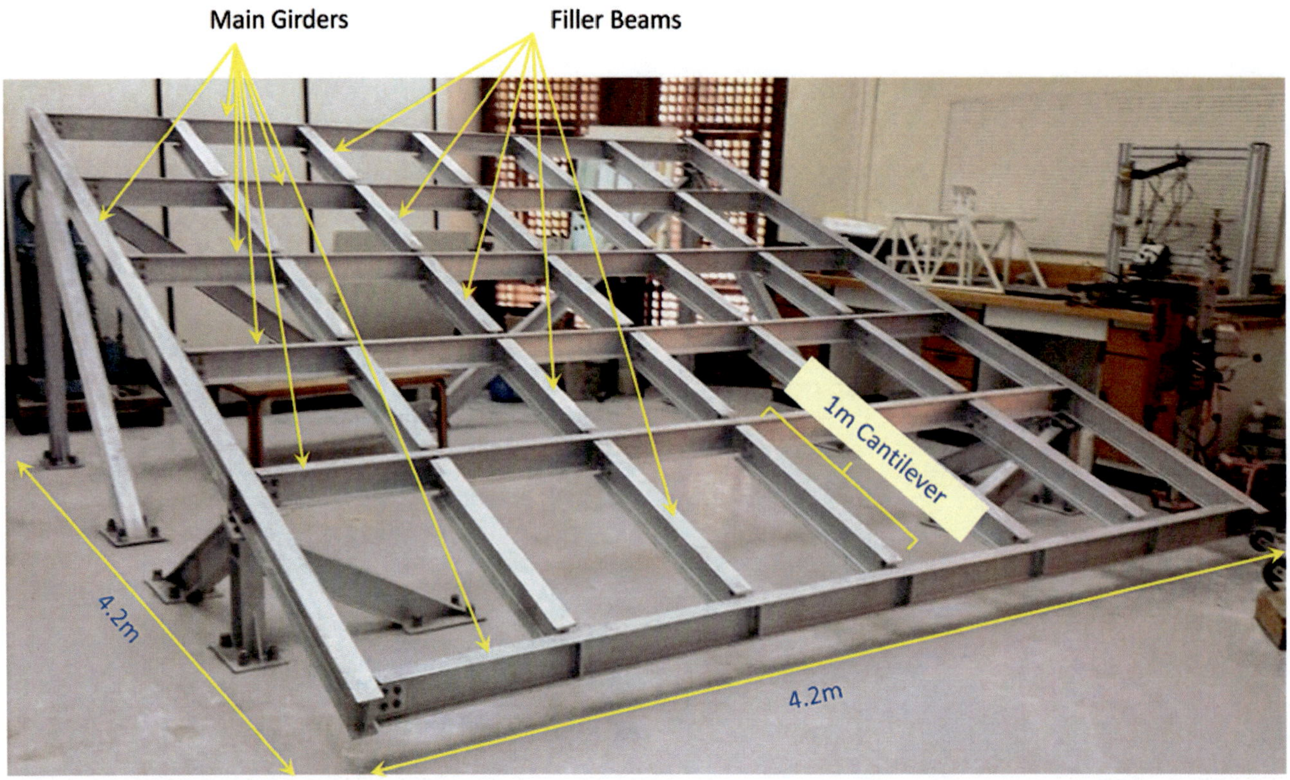

Fig. 1 The Benchmark structure

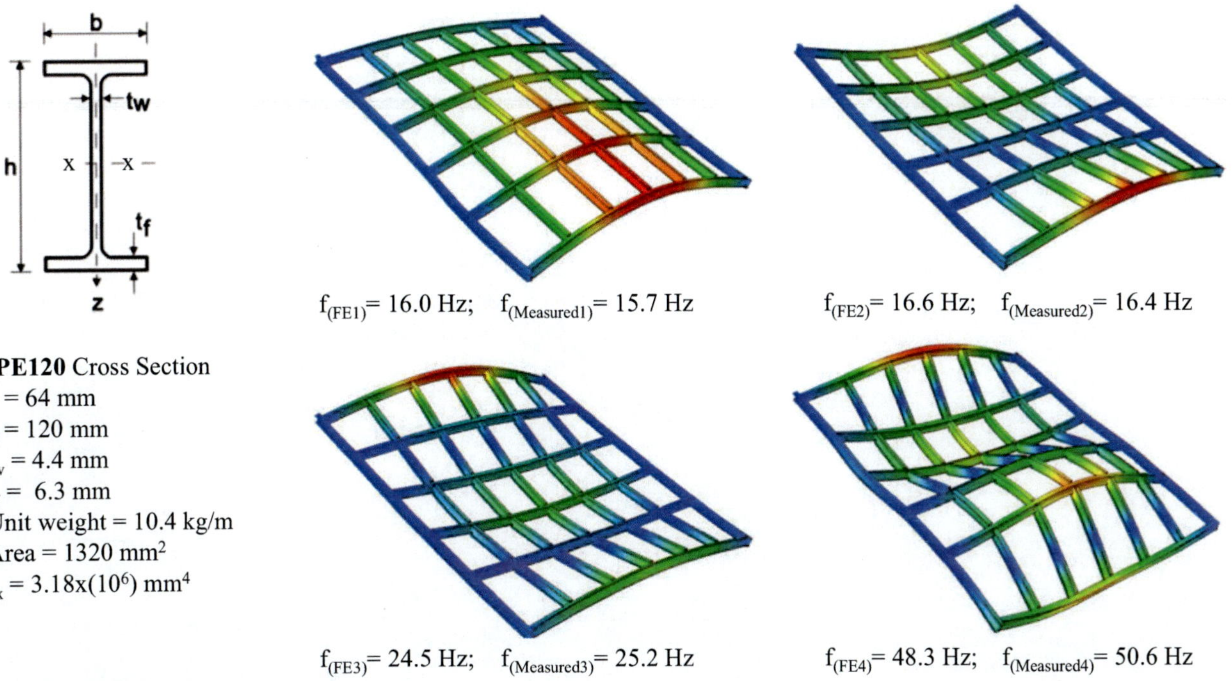

Fig. 2 IPE120 cross section and the first four bending modes

members are IPE120 cross sections (Fig. 2). Based on the structural drawings, a finite element model is created in Abaqus using C3D20R quadratic brick elements with 27 integration points (using 210 GPa for modulus of elasticity of steel), and the resulting first four governing bending mode shapes and corresponding modal frequencies are determined (Fig. 2). The measured frequencies per modal testing are processed in MEScope software and the values are shown in Fig. 2.

The total number of joints where the filler beams are being connected to the main girders is 30 (Fig. 3). One accelerometer was assigned for each joint and data was collected under various dynamic excitations without moving accelerometers between tests. Brüel & Kjær (model 8344) and PCB (model 393B04) accelerometers were used to record acceleration data, along with other dynamic testing equipment such as SmartAmp 2100E21-400 power amplifier, a dual 16-channel data acquisition system (DT9857E-16), and a modal shaker (Model 2100E11). Based on the type of dynamic excitation, a modal impact hammer or a shaker was utilized to create dynamic excitations as inputs.

For structural damage, the reference condition was considered as the undamaged state when all bolts are fully tightened (Fig. 4). Thirty additional conditions were individual bolt loosening at each joint shown in Fig. 3 (corresponding to 30 independent damage conditions). Loosening the bolts might be considered as a very slight damage, may be no damage at all. The authors wanted to consider bolt loosening as a damage based on the idea that when the damage detection methodology is sensitive enough to recognize and locate this slight damage condition and quantify it, this would be an indication of success for the methodology.

2 The Benchmark Website

The Benchmark dataset includes accelerations for each damage condition. The website address for the Benchmark dataset is:

- http://www.structuralvibration.com/benchmark/

  The main page of the Benchmark website includes links for additional information on the overview of the dataset package; grandstand simulator; instrumentation; damage scenarios; dataset descriptions; and downloading links (Fig. 5a). Users can navigate through the menu options on the left side of the main window. In addition to the Benchmark dataset, the damage detection tool used by the authors, One-Dimensional Convolutional Neural Networks (1D-CNNs) application codes and accompanied files are shared on this website:
- http://www.structuralvibration.com/cnns/

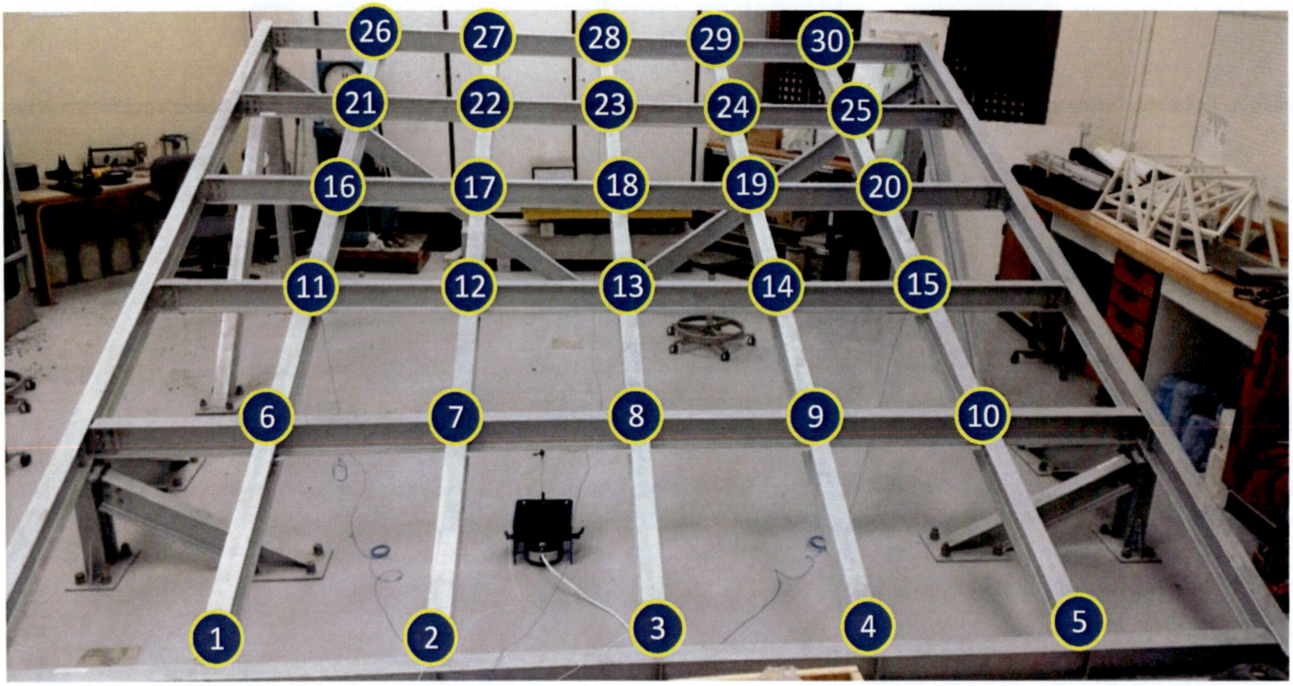

Fig. 3 Joint numbering on the Benchmark structure

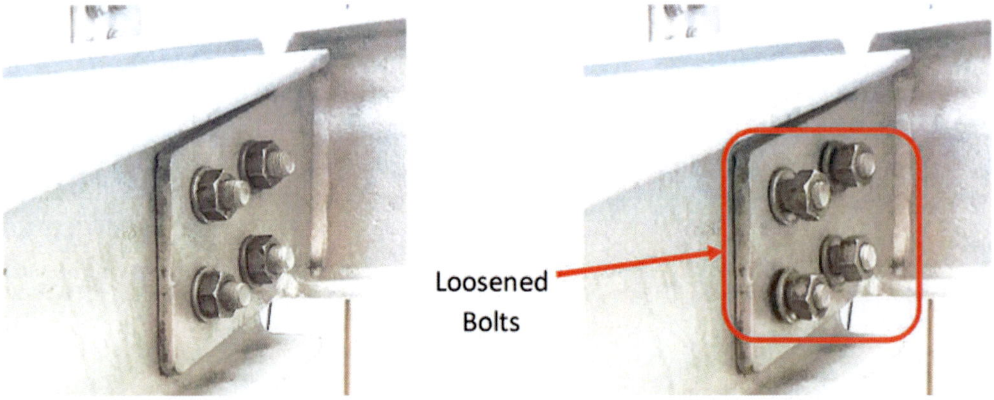

Fig. 4 The "damage" is introduced to the benchmark structure by loosening the bolts at the joints

Similarly, through the menu on the left side of the main window, the users can access information on back-propagation, and 1D-CNN package which includes the C++ Windows Console application; Matlab code; CDF files; configuration file; and downloading links (Fig. 5b).

After successful use of the original 2D version applications of CNNs [39–47], the 1D-CNNs were also proven to be an accurate and reliable tool used in various fields [48–51]. They are implemented as a damage identification/localization tool with the built-in adaptive architecture, merging the steps of feature extraction and classification into one compact unit. The first civil/structural engineering application of 1D-CNNs was conducted on the Benchmark structure, by the authors. The procedure runs directly on the raw acceleration signal, making the methodology available for real-time use.

The authors have conducted extensive analytical and experimental research on the Benchmark structure, mostly focusing on damage detection methods. Some of the publications produced based on the work on the Benchmark structure are [52–56], followed by a patent on the 1D-CNN SDD methodology [57]. Based on the success of the 1D-CNNs on the Benchmark structure as a damage detection and localization tool, the authors tested 1D-CNNs on other structures. For example, on another benchmark structure [58], 1D-CNNs showed excellent performance on identifying and quantifying the damage [59, 60]. In addition, the 1D-CNN code was also successful on damage identification on bearings as a rotating machine element [61, 62].

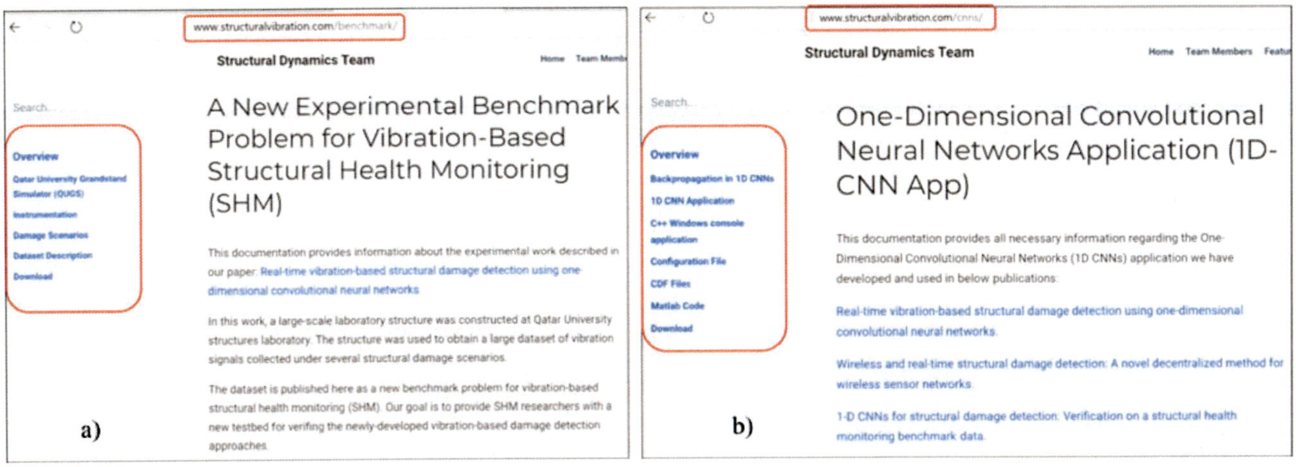

Fig. 5 Website views for (**a**) Benchmark dataset; (**b**) 1D-CNNs Application

3 Benchmark Dataset Description

The Benchmark dataset comprises two datasets to be downloaded (Dataset A and Dataset B). One dataset is for training and the other one is for testing purposes. The datasets include 31 TXT files (a total of 62 TXT files). Among the 31 files in one dataset, 30 files are dedicated to the damaged (untightened bolts) condition for each joint shown in Fig. 3. One additional TXT file is for the undamaged condition of the Benchmark structure. For each damaged condition, the accelerations were recorded for 256 s at each joint subjected to a white noise shaker excitation at a sampling frequency of 1024 Hz. This results in signals containing 256 × 1024 = 262,144 samples. For one dataset, this procedure was repeated 31 times, therefore resulting in 31 TXT files. The data collection process was done twice, resulting in two different datasets: Dataset A and Dataset B.

The naming of the TXT files is explained below:

Dataset A
- zzzAU.TXT: Undamaged condition
- zzzAD1.TXT: Damage at Joint 1
- zzzAD2.TXT: Damage at Joint 2
- ...
- zzzAD30.TXT: Damage at Joint 30

Dataset B
- zzzBU.TXT: Undamaged condition
- zzzBD1.TXT: Damage at Joint 1
- zzzBD2.TXT: Damage at Joint 2
- ...
- zzzBD30.TXT: Damage at Joint 30

Each TXT file contains 31 columns
- Column 1: timestamp.
- Column 2: Signal measured at Joint 1.
- Column 3: Signal measured at Joint 2.
- ...
- Column 31: Signal measured at Joint 30.

4 Conclusions

A new experimental Benchmark dataset generated from dynamic tests on a large-scale laboratory structure is presented in this paper. The Benchmark structure is a large-scale steel grid structure with footprint dimensions of 4.2 m $\times$ 4.2 m. It was constructed in a laboratory and has been used as a test bed for several publications by the authors. All structural members are hot-rolled steel IPE120 cross sections. For the structural damage detection studies, the damage was introduced at the joints of beam-to-girder connections by loosening the bolts. Even though bolt loosening is a slight change of rotational stiffness at the joints of the steel structure, the authors were able to detect, locate, and quantify it in their studies. The authors posted the dataset for 31 structural damage conditions on a public website as a new benchmark problem for structural damage detection at http://www.structuralvibration.com/benchmark/. The authors also shared the damage detection tool they used: One-Dimensional Convolutional Neural Networks (1D-CNNs). The 1D-CNNs package (application codes and accompanied files) is posted at http://www.structuralvibration.com/cnns/. The authors would appreciate feedback from other researchers on the Benchmark data and the 1D-CNNs package.

References

1. Agdas, D., Rice, J.A., Martinez, J.R., Lasa, I.R.: Comparison of visual inspection and structural-health monitoring as bridge condition assessment methods. J. Perform. Constr. Facil. (2016). https://doi.org/10.1061/(asce)cf.1943-5509.0000802
2. Gattulli, V., Chiaramonte, L.: Condition assesment by visual inspection for a bridge management system. Comput. Civ. Infrastruct. Eng. (2005). https://doi.org/10.1111/j.1467-8667.2005.00379.x
3. Živanović, S., Pavic, A., Reynolds, P.: Finite element modelling and updating of a lively footbridge: the complete process. J. Sound Vib. **301**, 126–145 (2007). https://doi.org/10.1016/j.jsv.2006.09.024
4. Zarei, J., Tajeddini, M.A., Karimi, H.R.: Vibration analysis for bearing fault detection and classification using an intelligent filter. Mechatronics. **24**, 151–157 (2014). https://doi.org/10.1016/j.mechatronics.2014.01.003
5. Yan, R., Gao, R.X., Chen, X.: Wavelets for fault diagnosis of rotary machines: a review with applications. Signal Process. **96**, 1–15 (2014). https://doi.org/10.1016/j.sigpro.2013.04.015
6. Raghavan, A., Cesnik, C.E.S.: Review of guided-wave structural health monitoring. Shock Vib. Dig. (2007). https://doi.org/10.1177/0583102406075428
7. Pavic, A., Reynolds, P., Prichard, S., Lovell, M.: Evaluation of mathematical models for predicting walking-induced vibrations of high-frequency floors. Int. J. Struct. Stab. Dyn. **03**, 107–130 (2003). https://doi.org/10.1142/S0219455403000756
8. Pacas, M., Villwock, S., Dietrich, R.: Bearing damage detection in permanent magnet synchronous machines. In: 2009 IEEE Energy Convers. Congr. Expo., pp. 1098–1103. IEEE, New York (2009). https://doi.org/10.1109/ECCE.2009.5316091
9. Mao, Z., Todd, M.D.: A Bayesian recursive framework for ball-bearing damage classification in rotating machinery. Struct. Health Monit. **15**, 668–684 (2016). https://doi.org/10.1177/1475921716656123
10. Díaz, I.M., Pereira, E., Reynolds, P.: Integral resonant control scheme for cancelling human-induced vibrations in light-weight pedestrian structures. Struct. Control Health Monit. (2012). https://doi.org/10.1002/stc.423
11. Chen, H.-P., Ni, Y.-Q.: Structural Health Monitoring of Large Civil Engineering Structures. John Wiley & Sons, Ltd, Chichester, UK (2018). https://doi.org/10.1002/9781119166641
12. Royvaran, M., Avci, O., Davis, B.: An overview on floor vibration serviceability evaluation methods with a large database of recorded floor data. In: Conf. Proc. Soc. Exp. Mech. Ser. (2021). https://doi.org/10.1007/978-3-030-47634-2_10
13. Ngoan, D.T., Mustafa, G., Osama, A., Onur, A.: Stadium vibration assessment for serviceability considering the vibration duration. In: Proceedings, Annu. Conf. - Can. Soc. Civ. Eng. (2017)
14. Muhammad, Z., Reynolds, P., Avci, O., Hussein, M.: Review of pedestrian load models for vibration serviceability assessment of floor structures. Vibration. (2018). https://doi.org/10.3390/vibration2010001
15. Chaabane, M., Ben Hamida, A., Mansouri, M., Nounou, H.N., Avci, O.: Damage detection using enhanced multivariate statistical process control technique. In: 2016 17th Int. Conf. Sci. Tech. Autom. Control Comput. Eng. STA 2016 – Proc. (2017). https://doi.org/10.1109/STA.2016.7952052
16. Celik, O., Do, N.T., Abdeljaber, O., Gul, M., Avci, O., Catbas, F.N.: Recent issues on stadium monitoring and serviceability: a review. In: Conf. Proc. Soc. Exp. Mech. Ser. (2016). https://doi.org/10.1007/978-3-319-29763-7_41
17. Celik, O., Catbas, F.N., Do, N.T., Gul, M., Abdeljaber, O., Younis, A., Avci, O.: Issues, codes and basic studies for stadium dynamics. In: Proc. Second Int. Conf. Infrastruct. Manag. Assess. Rehabil. Tech., Sharjah, UAE (2016)
18. Catbas, F.N., Celik, O., Avci, O., Abdeljaber, O., Gul, M., Do, N.T.: Sensing and monitoring for stadium structures: a review of recent advances and a forward look. Front. Built Environ. **3**, 38 (2017). https://doi.org/10.3389/fbuil.2017.00038
19. Alabbasi, S., Hussein, M., Abdeljaber, O., Avci, O.: A numerical and experimental investigation of a special type of floating-slab tracks. Eng. Struct. (2020). https://doi.org/10.1016/j.engstruct.2020.110734
20. Alabbasi, S., Hussein, M., Abdeljaber, O., Avci, O.: Investigating the dynamics of a special type of a floating-slab tracks. In: COMPDYN Proc. (2019). https://doi.org/10.7712/120119.6962.19626
21. Carden, E.P., Fanning, P.: Vibration based condition monitoring: a review. Struct. Health Monit. (2004). https://doi.org/10.1177/1475921704047500

22. Abdeljaber, O., Avci, O., Do, N.T., Gul, M., Celik, O., Necati Catbas, F.: Quantification of structural damage with self-organizing maps. Conf. Proc. Soc. Exp. Mech. Ser. (2016). https://doi.org/10.1007/978-3-319-29956-3_5
23. Y.Y. Li, Hypersensitivity of strain-based indicators for structural damage identification: a review, Mech. Syst. Signal Process. (2010). doi:https://doi.org/10.1016/j.ymssp.2009.11.002
24. Mansouri, M., Avci, O., Nounou, H., Nounou, M.: Iterated square root unscented Kalman filter for state estimation - CSTR model. In: 12th Int. Multi-Conference Syst. Signals Devices, SSD 2015 (2015). https://doi.org/10.1109/SSD.2015.7348243
25. Farrar, C.R., Worden, K.: Structural Health Monitoring: A Machine Learning Perspective. Wiley, Hoboken, NJ (2012). https://doi.org/10.1002/9781118443118
26. Quek, S.T., Tran, V.A., Hou, X.Y., Duan, W.H.: Structural damage detection using enhanced damage locating vector method with limited wireless sensors. J. Sound Vib. **328**, 411–427 (2009). https://doi.org/10.1016/j.jsv.2009.08.018
27. Mansouri, M., Avci, O., Nounou, H., Nounou, M.: Iterated square root unscented Kalman filter for nonlinear states and parameters estimation: three DOF damped system. J. Civ. Struct. Health Monit. **5** (2015). https://doi.org/10.1007/s13349-015-0134-7
28. Ghahari, S.F., Abazarsa, F., Avci, O., Çelebi, M., Taciroglu, E.: Blind identification of the Millikan Library from earthquake data considering soil-structure interaction. Struct. Control Health Monit. (2016). https://doi.org/10.1002/stc.1803
29. Mansouri, M., Avci, O., Nounou, H., Nounou, M.: A comparative assessment of nonlinear state estimation methods for structural health monitoring. In: Conf. Proc. Soc. Exp. Mech. Ser. (2015). https://doi.org/10.1007/978-3-319-15224-0_5
30. Ghahramani, Z.: Probabilistic machine learning and artificial intelligence. Nature. (2015). https://doi.org/10.1038/nature14541
31. Kiranyaz, S., Avci, O., Abdeljaber, O., Ince, T., Gabbouj, M., Inman, D.J.: 1D convolutional neural networks and applications: a survey. Mech. Syst. Signal Process. **151** (2021). https://doi.org/10.1016/j.ymssp.2020.107398
32. Avci, O., Abdeljaber, O., Kiranyaz, S., Hussein, M., Gabbouj, M., Inman, D.J.: A review of vibration-based damage detection in civil structures: from traditional methods to machine learning and deep learning applications. Mech. Syst. Signal Process. (2021). https://doi.org/10.1016/j.ymssp.2020.107077
33. Kubat, M.: An Introduction to Machine Learning. Springer, New York (2017). https://doi.org/10.1007/978-3-319-63913-0
34. Rafiei, M.H., Adeli, H.: A novel machine learning-based algorithm to detect damage in high-rise building structures. Struct. Des. Tall Spec. Build. (2017). https://doi.org/10.1002/tal.1400
35. Avci, O., Abdeljaber, O., Kiranyaz, S., Inman, D.: Control of plate vibrations with artificial neural networks and piezoelectricity. In: Conf. Proc. Soc. Exp. Mech. Ser. (2020). https://doi.org/10.1007/978-3-030-12676-6_26
36. Abdeljaber, O., Avci, O., Inman, D.J.: Active vibration control of flexible cantilever plates using piezoelectric materials and artificial neural networks. J. Sound Vib. **363** (2016). https://doi.org/10.1016/j.jsv.2015.10.029
37. Avci, O., Abdeljaber, O., Kiranyaz, S.: Structural damage detection in civil engineering with machine-learning: current state of the art. In: Conf. Proc. Soc. Exp. Mech. Ser. (2021)
38. Avci, O., Abdeljaber, O., Kiranyaz, S.: An overview of deep learning methods used in vibration-based damage detection in civil engineering. In: Conf. Proc. Soc. Exp. Mech. Ser. (2021)
39. Dai, J., Qi, H., Xiong, Y., Li, Y., Zhang, G., Hu, H., Wei, Y.: Deformable convolutional networks. In: Proc. IEEE Int. Conf. Comput. Vis. (2017). https://doi.org/10.1109/ICCV.2017.89
40. Milosevic, N.: Introduction to Convolutional Neural Networks. Apress, New York (2020). https://doi.org/10.1007/978-1-4842-5648-0
41. Tan, M., Le, Q.V.: EfficientNet: rethinking model scaling for convolutional neural networks. In: 36th Int. Conf. Mach. Learn. ICML 2019 (2019)
42. Niepert, M., Ahmad, M., Kutzkov, K.: Learning convolutional neural networks for graphs. In: 33rd Int. Conf. Mach. Learn. ICML 2016 (2016)
43. Zhang, Y., Gao, J., Zhou, H.: Breeds classification with deep convolutional neural network. In: ACM Int. Conf. Proceeding Ser. (2020). https://doi.org/10.1145/3383972.3383975
44. Kalash, M., Rochan, M., Mohammed, N., Bruce, N.D.B., Wang, Y., Iqbal, F.: Malware classification with deep convolutional neural networks. In: 2018 9th IFIP Int. Conf. New Technol. Mobil. Secur. NTMS 2018 – Proc. (2018). https://doi.org/10.1109/NTMS.2018.8328749
45. Lopez Pinaya, W.H., Vieira, S., Garcia-Dias, R., Mechelli, A.: Convolutional neural networks. In: Mach. Learn. Methods Appl. to Brain Disord. (2019). https://doi.org/10.1016/B978-0-12-815739-8.00010-9
46. Kim, Y.: Convolutional neural networks for sentence classification. In: EMNLP 2014–2014 Conf. Empir. Methods Nat. Lang. Process. Proc. Conf. (2014). https://doi.org/10.3115/v1/d14-1181
47. Krizhevsky, A., Sutskever, I., Hinton, G.E.: ImageNet classification with deep convolutional neural networks. Commun. ACM. (2017). https://doi.org/10.1145/3065386
48. Ince, T., Kiranyaz, S., Eren, L., Askar, M., Gabbouj, M.: Real-time motor fault detection by 1-D convolutional neural networks. IEEE Trans. Ind. Electron. (2016). https://doi.org/10.1109/TIE.2016.2582729
49. Kiranyaz, S., Ince, T., Gabbouj, M.: Real-time patient-specific ECG classification by 1-D convolutional neural networks. IEEE Trans. Biomed. Eng. **63**, 664–675 (2016). https://doi.org/10.1109/TBME.2015.2468589
50. Kiranyaz, S., Gastli, A., Ben-Brahim, L., Alemadi, N., Gabbouj, M.: Real-time fault detection and identification for MMC using 1D convolutional neural networks. IEEE Trans. Ind. Electron. (2018). https://doi.org/10.1109/TIE.2018.2833045
51. Kiranyaz, S., Ince, T., Gabbouj, M.: Personalized monitoring and advance warning system for cardiac arrhythmias. Sci. Rep. **7** (2017). https://doi.org/10.1038/s41598-017-09544-z
52. Abdeljaber, O., Avci, O., Kiranyaz, S., Gabbouj, M., Inman, D.J.: Real-time vibration-based structural damage detection using one-dimensional convolutional neural networks. J. Sound Vib. **388**, 154–170 (2017). https://doi.org/10.1016/j.jsv.2016.10.043
53. Abdeljaber, O., Younis, A., Avci, O., Catbas, N., Gul, M., Celik, O., Zhang, H.: Dynamic testing of a laboratory stadium structure. In: Geotech. Struct. Eng. Congr. 2016 - Proc. Jt. Geotech. Struct. Eng. Congr. 2016 (2016). https://doi.org/10.1061/9780784479742.147
54. Avci, O., Abdeljaber, O., Kiranyaz, S., Inman, D.: Structural damage detection in real time: implementation of 1D convolutional neural networks for SHM applications. In: Niezrecki, C. (ed.) Struct. Heal. Monit. Damage Detect. Vol. 7 Proc. 35th IMAC, A Conf. Expo. Struct. Dyn. 2017, pp. 49–54. Springer International Publishing, Cham (2017). https://doi.org/10.1007/978-3-319-54109-9_6
55. Avci, O., Abdeljaber, O., Kiranyaz, S., Hussein, M., Inman, D.J.: Wireless and real-time structural damage detection: a novel decentralized method for wireless sensor networks. J. Sound Vib. (2018)

56. Avci, O., Abdeljaber, O., Kiranyaz, S., Inman, D.: Convolutional neural networks for real-time and wireless damage detection. In: Conf. Proc. Soc. Exp. Mech. Ser. (2020). https://doi.org/10.1007/978-3-030-12115-0_17
57. Kiranyaz, S., Avci, O., Jaber, O.A.Q.A.: Real-time structural damage detection by convolutional neural networks. US16031519. https://patents.google.com/patent/US20190017911A1/en (2019)
58. Dyke, S.J., Bernal, D., Beck, J., Ventura, C.: Experimental phase II of the structural health monitoring benchmark problem. In: Proc. 16th ASCE Eng. Mech. Conf. (2003)
59. Abdeljaber, O., Avci, O., Kiranyaz, M.S., Boashash, B., Sodano, H., Inman, D.J.: 1-D CNNs for structural damage detection: verification on a structural health monitoring benchmark data. Neurocomputing. (2017). https://doi.org/10.1016/j.neucom.2017.09.069
60. Avci, O., Abdeljaber, O., Kiranyaz, M.S., Boashash, B., Sodano, H., Inman, D.J.: Efficiency validation of one dimensional convolutional neural networks for structural damage detection using a SHM benchmark data. In: 25th Int. Congr. Sound Vib. 2018, ICSV 2018 Hiroshima Call., p. 2018
61. Abdeljaber, O., Sassi, S., Avci, O., Kiranyaz, S., Ibrahim, A.A., Gabbouj, M.: Fault detection and severity identification of ball bearings by online condition monitoring. IEEE Trans. Ind. Electron. (2019). https://doi.org/10.1109/TIE.2018.2886789
62. Avci, O., Abdeljaber, O., Kiranyaz, S., Sassi, S., Ibrahim, A., Gabbouj, M.: One dimensional convolutional neural networks for real-time damage detection of rotating machinery. In: Conf. Proc. Soc. Exp. Mech. Ser. (2021)

Implementation of an Organic Database Structure for Population-Based Structural Health Monitoring

Daniel S. Brennan, Chandula T. Wickramarachchi, Elizabeth J. Cross, and Keith Worden

Abstract Population-based Structural Health Monitoring (PBSHM) considers grouping multiple structures into a single population, thus allowing additional insight to be gained from SHM feature data as a whole, compared to insights gained for any individual structure. Current data storage methodologies for SHM often operate from a single structure point of view. Different types of data can often be stored in different systems, files, structures and languages. This approach means that bespoke methodologies must be developed for any new SHM application, making analysing and processing of SHM data costly. The advantage of the proposed unified approach is the facilitation of information sharing across populations of structures, a key principle in PBSHM. A generic pool of methodologies can then be developed, allowing any set of PBSHM data to be quickly and efficiently analysed and processed, bringing down the total cost of running a system.

In this paper a PBSHM schema is defined to store multiple populations of PBSHM data inside a NoSQL database via a *Time First* approach. The schema allows a single database that organically grows as more data/knowledge about the population becomes available. As everything is grouped by Time, the approach allows previously unknown data relationships to be easily viewed. In this paper the PBSHM schema is implemented in a MongoDB database and sensor data are added into the system to show the ease of use of a standardised data schema.

Keywords Population-based structural health monitoring · PBSHM · Database · MongoDB · NoSQL · Schema

1 Introduction

When dealing with data in a traditional Structural Health Monitoring (SHM) scenario, one would generally only consider a single structure and subsequent data set [1]. In Population-based Structural Health Monitoring (PBSHM) [2, 3], multiple structures of different types are monitored to gain additional insights as a whole compared to the knowledge gained from any single structure. However, there are inherent issues with this approach; each structure comes with its own set of labels, units and formats, which produces significant problems when comparing data.

There are existing standards which address these data requirements for a particular type of structure. Jeong et al. [4] discuss existing standards used within bridge lifecycle management, such as the Bridge Information Modeling (BrIM) standard. These standards provide a data structure, enabling easy comparison; however, they lack the functionality to extend the specification to include different structure types and associated data types/units. This issue often means bespoke methodologies must be developed for every type of structure within an SHM application; not only does it make analysing and processing of PBSHM data costly, it puts unnecessary barriers in place for adoption of PBSHM technologies.

Therefore, PBSHM requires a new standardised methodology towards data storage that addresses the unique challenge of accommodating multiple structure types and varying data requirements. This paper aims to explore current database technologies and proposes a new standard for the structuring of PBSHM data. The standard would facilitate a uniform way of storing data on structures, support the flexibility of different data types and enable easy identification of structures within a population or type. It is important to acknowledge that structure owners may have existing systems in place to monitor and capture SHM data; in order for industrial adoption and adaptation, the standard must work in conjunction with existing systems.

D. S. Brennan (✉) · C. T. Wickramarachchi · E. J. Cross · K. Worden
Dynamics Research Group, Department of Mechanical Engineering, University of Sheffield, Sheffield, UK
e-mail: dsbrennan1@sheffield.ac.uk; c.t.wickramarachchi@sheffield.ac.uk; e.j.cross@sheffield.ac.uk; k.worden@sheffield.ac.uk

K. Grimmelsman (ed.), *Dynamics of Civil Structures, Volume 2*, Conference Proceedings of the Society for Experimental Mechanics Series, https://doi.org/10.1007/978-3-030-77143-0_3

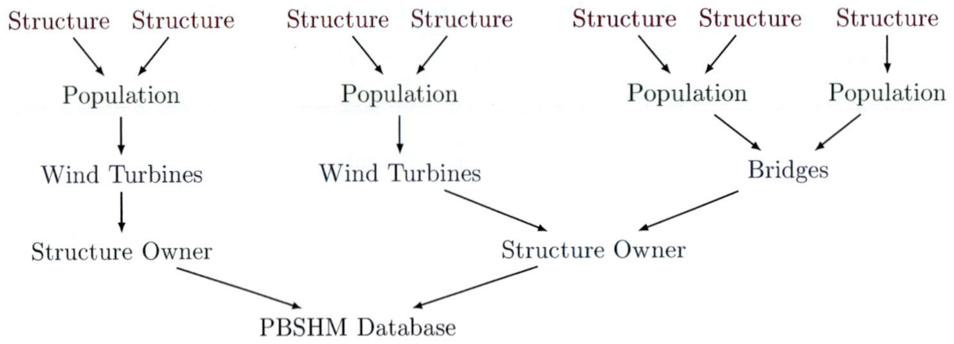

Fig. 1 Data transmission from structure owners' current SHM systems into a PBSHM database

There are two main methodologies for storing of data; flat files (such as Comma-Separated Values (CSV) and plain text) and databases. Whilst a flat file structure facilitates the viewing of data as a trivial task, querying and building relationships between these data can become difficult when comparing data across multiple files. In comparison, when using a database, querying and comparing data becomes easy whilst importing data can become a more difficult task, especially when importing large amounts of historic data. As a key principle of PBSHM is to facilitate increased knowledge retrieval, having a data store that allows data to be retrieved easily and efficiently is key.

Due to the requirement of working in conjunction with existing systems, the PBSHM database should allow structure owners to capture SHM data as their current business practices define. Afterwards, data should be submitted into a centralised PBSHM database. The PBSHM database needs to have a standardised format not only for data storage but also one for data transmission. The PBSHM database must allow operation in the following scenarios: an open global database that structure owners can submit data into, or a private internal database allowing an owner to compare data across their assets (e.g. Wind and Wave farms). Whichever route a structure owner may decide, the format of how to store and communicate data must be the same (see Fig. 1).

This paper outlines the creation of a PBSHM *schema* to govern data permitted inside the PBSHM database; Sect. 2 outlines the features a PBSHM schema would require; Sect. 3 evaluates current database technology against the schema requirements and Sect. 4 explores the design process for creation of the proposed PBSHM schema.

2 PBSHM Schema Requirements

A schema is to a database what an architect's drawing is to a structure. It is a technical document that says where data should go, what is allowed where and outlines any relationships and hierarchical structure of the data. There are several considerations that must be taken into account throughout the design of this schema:

- PBSHM is constantly evolving and, as such, has new data requirements that are not yet known. When adding unknown data into the database, any existing data inside the system need to remain valid.
- There may be vast amounts of historical SHM data that need to be imported into the system. Historical data may not be complete or have all the data the schema can hold. Subsequently, the schema should group different types of data together and allow each group to be optional.
- Engineers from different fields will be consuming this data, thus the schema must be straightforward to understand and master. The schema needs not to become another barrier for the adoption of PBSHM technologies.

In short, the data structure that the schema governs must be an organic data structure by design. Organic data structures are designed and engineered from the ground up, with the knowledge that there is only a partial understanding of data requirements. This design methodology for data structures allows current knowledge of data requirements to be included at the outset; however, the design leaves space for additional data to be defined afterwards. It is imperative to achieve this without invalidating or changing the meaning and context of existing data in the database. Subsequently, the database can grow or shrink over time as requirements of structural monitoring change.

Once a standardised structure of data exists via the PBSHM schema, it facilitates the development of generic algorithms and methodologies to work seamlessly across data sets inside PBSHM. Wickramarachchi et al. [5] take this idea further, with the introduction of modules to encapsulate these generic algorithms and methodologies to create a PBSHM framework. At

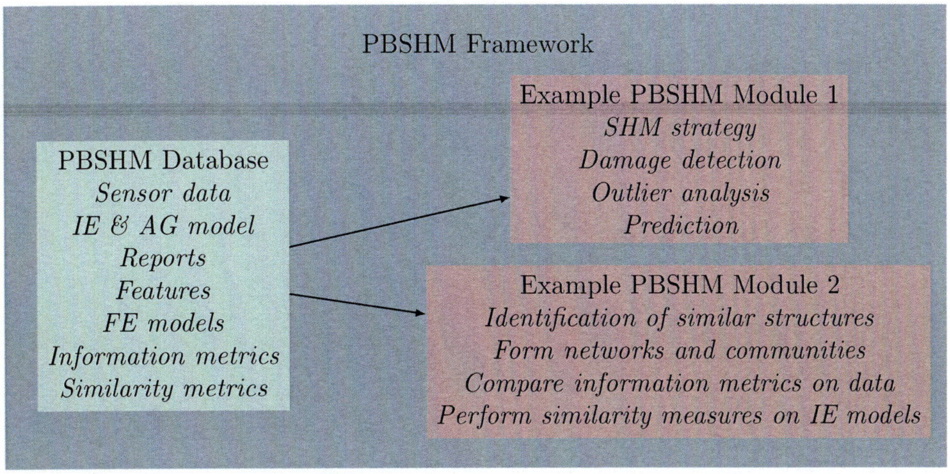

Fig. 2 PBSHM Framework (recreated from Wickramarachchi et al. [5]) interacting with the PBSHM database

the core of the PBSHM Framework is the PBSHM database governed by the PBSHM Schema, with PBSHM modules on top of this.

Wickramarachchi et al. [5] lay out the foundations of what may be stored within a PBSHM database (see Fig. 2). The current paper will focus on incorporating the sensor data requirement outlined by Wickramarachchi et al. [5] into the PBSHM Schema.

3 Database

Databases, at their core, are simply a technique for storing data in a structured manner. Codd [6] introduced this idea in 1970 and it has since become the foundation of modern *Relational database management systems* (RDBMS). However, since the introduction of this principle, the constraint of data being structured/relational has caused a new breed of databases to be born. *Not only Structured Query Language* (NoSQL) are a family of database methodologies that aim to solve the structured data problem by enabling schema-free data storage. *Structured Query Language* (SQL) is used to manage data inside an RDBMS, and as such, NoSQL databases are designed without these inherent constraints. There are *vertical* and *horizontal* scalability concerns due to the nature of an RDBMS. *Vertical* and *horizontal* scaling are the principle of how to facilitate the required growth of a system; *vertical* views scaling from a single item/node, *horizontal* views scaling from multiple items/nodes. Nance et al. [7] and Zafar et al. [8] review these concerns in conjunction with detailing differences between RDBMS and NoSQL. Gandini et al. [9] outline some performance-planning considerations when using a NoSQL database.

It should be noted that some of the security concerns highlighted by Nance et al. [7] regarding encryption at rest,[1] authentication in a *sharded*[2] environment and client communication encryption have been addressed since publication [10–13].

Selecting the correct database technology is crucial to the adoption of the PBSHM schema and database. Subsequently, each database type is required to be evaluated against the PBSHM requirements set forth in this paper.

3.1 RDBMS

RDBMS, in its most simplified version, stores data in the format of tables and columns (see Fig. 3). Relationships within are created via *primary keys* and *foreign keys*. Primary keys are the method for identifying a unique data entry inside a table. Foreign keys are the method for creating a link/reference to a column in another table, usually a primary key; thus

[1]Encrypted Storage Engines are available as of MongoDB 3.2 Enterprise [10, 11].

[2]*Sharding* is a method of splitting data across multiple database instances.

Structure table

id	name	type
1	structure-1	turbine

Structure Sensor table

id	structure id	sensor name
1	1	sensor-1
2	1	sensor-2
3	1	sensor-3
4	1	sensor-4

Fig. 3 An relational example showing a 'structure-1' structure containing four sensors within an RDBMS

```
{
    "id":1,
    "name":"structure-1",
    "sensors":[
        "sensor-1","sensor-2","sensor-3","sensor-4"
    ]
}
```

Fig. 4 An embedded example showing a 'structure-1' structure containing four sensors within a NoSQL Document database

creating a relationship between the tables. Utilising the combination of primary keys and foreign keys facilitates the retrieval of relational data. To encourage separation of data into multiple tables for the purpose of data deduplication, RDBMS have a normalisation concept.

Database normalisation provides a methodology for when to partition data off into a separate table by comparing the data inside the structure to the normal forms; consequently, it structures data in an approach that grants extensions to the structure, without the necessity of invalidating existing data. Codd [6] introduced what is now known as first normal form (1NF) and later introduced second and third normal form (2NF and 3NF) [14, 15]. Each level of normal form adds additional requirements to reduce data duplication and improve data integrity; the greater number of columns that contain identical data regarding an entity, the greater number of columns a system is required to update when data change.

The benefit of an RDBMS is proven system stability; however, they lack the flexibility to enable yet unknown data to be added into the database without a schema change. It is possible to design around this issue in a data schema, nevertheless it is not recommended, due to the required complexity of said schema. In the use case of PBSHM, multiple SQL statements are required to retrieve the state of a single structure.

3.2 *NoSQL*

NoSQL databases are generally agreed to be divided into four sub-categories;

- Key-Value databases;
- Wide-Column databases;
- Document databases;
- Graph databases.

Key-Value databases are, at their core, a dictionary data structure; they are a large collection of key-value pairs, requiring each key to be unique. Wide-Column databases are similar to RDBMS; they use tables, columns and rows. However, each row inside the database may have differentiating columns compared to sibling rows; similar to a 'dictionary of dictionaries' data structure. Graph databases use nodes and edges to store data; they function via the same rules as graph theory. Document databases use the notion of an entry being a document; they follow a similar structure to *Object Oriented Programming* (OOP). Documents have multiple properties/attributes; however, they can also have nested documents within themselves. Zafar et al. [8] discuss the merits and drawbacks of each type of NoSQL database in additional detail.

Within the context of a PBSHM database, Document databases permit the functionality of both enabling unknown data to be added into the database without prior knowledge and enabling nesting of data within one document (see Fig. 4). There is no feature to store relationships between documents; however, in the case of PBSHM databases this is not a problem, given that any relationship between structures would need to be discovered via algorithms inside the PBSHM framework first. Document databases provide a unique opportunity for all relevant information regarding the state of a structure to be returned with one document, thus facilitating a straightforward and simple schema for engineers to understand.

3.3 Database Choice

Both RDBMS and NoSQL Document databases provide the technical requirements to implement a PBSHM database. However, with an RDBMS there are additional required processes due to converting tabular data into object representations. Consequently, NoSQL document databases have been chosen here for implementation of the PBSHM database.

There are multiple NoSQL document databases available, each providing a unique set of features driven by community goals. When selecting a document database, the following additional requirements were introduced: access controls to enable different structure permissions, simplistic document structure and document level concurrency. For the purposes of the PBSHM database, MongoDB has been chosen. MongoDB was initially developed in 2007 [16], and has since become a widely adopted and supported document database. MongoDB can be deployed on premises, on cloud providers or via MongoDB Atlas.

Documents inside MongoDB are stored in Binary JavaScript Object Notation (BSON) format—a derivative of the JavaScript Object Notation (JSON) format. JSON [17, 18] is a text-based data format that has gained popularity due to its simplistic document structure, lightweight footprint and self-contained nature.

MongoDB uses the following levels for data storage:

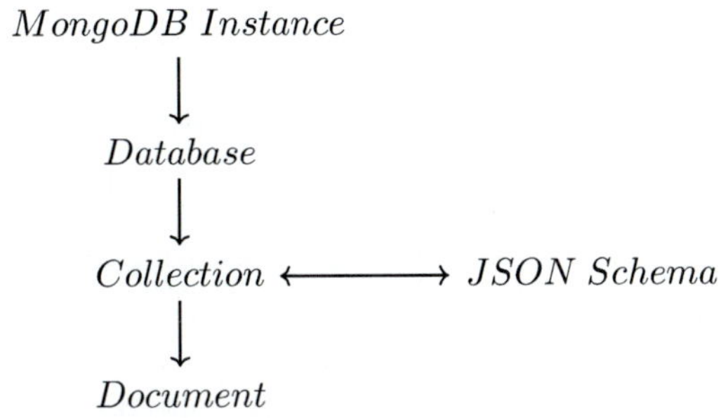

An instance of MongoDB can contain multiple databases; inside each database can be multiple collections and inside a collection can be multiple documents. Whilst a principle of NoSQL document databases is a schema-less design, MongoDB supports document validation via associating each collection with a JSON Schema. This facilitates a 'pbshm' database inside the MongoDB instance, a global 'structures' collection inside the database using the PBSHM JSON Schema for document validation and structure data documents inside the collection.

Due to MongoDB's access controls [10, 19], all users will be given read permission to the 'structures' collection; if any PBSHM framework modules require storage of computed values, these will be stored in module specific collections. The MongoDB database here will use the WiredTiger [10, 20, 21] storage engine, which supports document level concurrency, thus facilitating a structure's data to be updated on the system without locking read access to the rest of the structures.

Facilitating minimal latency upon retrieval of common queried data is implemented via *indexes* [10, 22]. Indexes are quintessentially pointers from a specific data value to documents which are stored in memory. For example, an index could be created on a population name, thus when the system is queried for documents inside a population, the system already knows which corresponding documents require to be returned without enumerating through each document inside a collection.

Distribution of data, either for redundancy or scalability is a key component for any database. MongoDB supports this via *replication* [10, 23] and *sharding* [10, 24]; however, for the purpose of this paper, they are not implemented within the PBSHM database due to the system only containing historical data that can fit within a single database node.

4 The PBSHM Schema

When designing a database schema, the underlying database software limitations and best practices must be taken into account. MongoDB outlines two different data models for structuring data within itself [10, 25]. *Normalised data models* [10, 26] store relationships to related documents via identifier reference fields, whereas *embedded data models* [10, 27] embed relational data as sub-documents within the parent document.

Embedded data models facilitate all related data being returned without requiring additional queries; however, data duplication is present due to the nature of the design. Given that understanding the data returned is a key requirement, embedded data models have been chosen for the PBSHM schema.

4.1 Organic Design

Fundamental to implementation of organic schema design, is enabling horizontal and vertical scaling with the schema. Commonly, this objective is achieved by splitting data into different objects/groups and placing these inside a hierarchical structure. Horizontal scaling is achieved by nesting an array of objects within the parent object under a property. Vertical scaling is achieved by adding additional properties or embedded objects into the existing objects. For instance, if one takes the example of PBSHM structure and channel data; these could be grouped into the following: Structure Details, Sensor Data and Time/Value readings. A non-organic example is below (see Appendix 1 for associated schema):

```
{
    "name":"structure 1",
    "channel1Name":"acceleration channel 1",
    "channel1Data":[],
    "channel2Name":"velocity channel 1",
    "channel2Data":[]
}
```

However, every time a new channel is required, the schema would need to be changed to support this. If the data are restructured to extract all channel data into their own object, an array of channel objects can then be placed beneath a channel property in the root document. An organic example is below (see Appendix 2 for associated schema):

```
{
    "name":"structure 1",
    "channels":[
        {"name":"channel 1","data":[]},
        {"name":"channel 2","data":[]}
    ]
}
```

Changing the structure of the document has enabled horizontal scaling for channel data. Additional channels may be included without prior knowledge of the number of channels, or requiring each structure to contain an equal number of channels. Additional sections may also be included into the data without invalidating the current data due to the nature of NoSQL document databases. For instance, if the report section from a PBSHM framework was included, the data could be stored under a report property in the root document. An organic example with an additional report section is below:

```
{
    "name":"structure 1",
    "channels":[
        {"name":"channel 1","data":[]},
        {"name":"channel 2","data":[]}
    ],
    "reports":[
        {"name":"channel 1","data":[]},
        {"name":"channel 2","data":[]}
    ]
}
```

The data within sub-objects get additional context from their position within the structure. For instance, one determines what context the *name* property has by its document path, if the path of *name* is *$.channels.name*, it is the name of the channel;

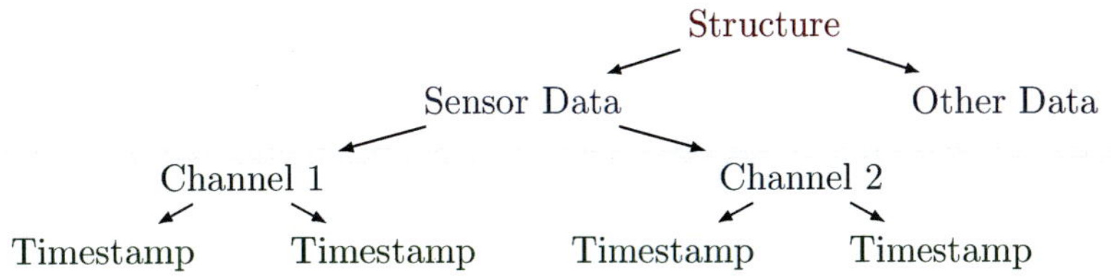

Fig. 5 Hierarchical structure in a Channel First approach

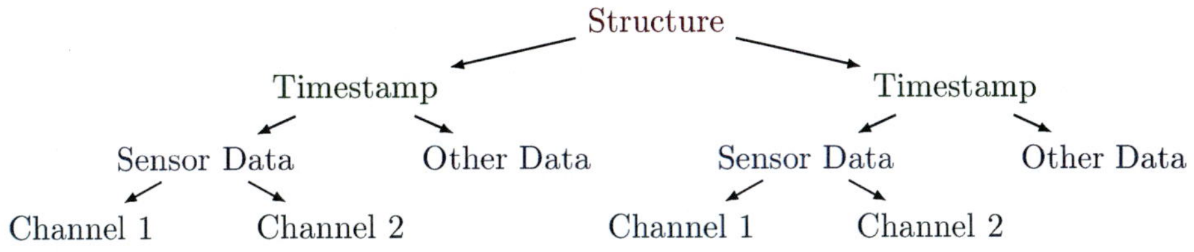

Fig. 6 Hierarchical structure in a Time First approach

however, if the path is *$.reports.name*, it is the name of a report. The schema has enabled organic data growth without implementation changes. However, if the requirements to validate data inside the report section are changed, the schema can still be updated to validate report data without invalidating sensor data.

4.2 *Time First Approach*

Traditionally, SHM data are laid out with the priority given to where the data came from. For instance, with sensor data, the data are often structured via Channel First, which provides a hierarchy of structure, sensor data, channel and timestamp (see Fig. 5). However, this paper proposes giving the highest priority in the hierarchy to when data occurred instead of where they came from; a *Time First* approach (see Fig. 6).

A *Time First* approach enables data unification across sources in the PBSHM database, as time is a common variable in each data source. This approach facilitates a document structure where each document represents a snapshot in time of the knowledge of a structure. Having all known data from one time point in a single document enables engineers to efficiently see what monitoring data are available without enumerating through multiple documents.

Data from a source should not be present in a Time First document if it does not exist. For instance, when a structure is under construction, sensors may be fitted as construction progresses, which allows simple representation of data for end users. Data can be captured from these sensors without entering in data for sensors that do not exist on the structure at the given point in time. The opposite is true for a damage scenario; if a sensor is placed onto a section of a structure which is removed due to failure, capture of any sensors on this section of the structure should not be present in the document.

MongoDB has a size limit of 16 MB per document without implementing additional file structures [28]. By using a Time First approach, this is naturally limiting the quantity of data within a single document. Splitting data into smaller documents also increases performance due to concurrency; WiredTiger places a lock on a *per document* basis; if there was one document with large amounts of data across multiple time periods, this would be locked for reading until any updates on the document were complete.

4.3 *Implementation*

For the PBSHM Schema, data are separated into individual nested entities (see Fig. 7). Each entity corresponds to a relevant source from the channel data capture process; the structure itself (*Structure*), the channels monitoring the structure (*Channel*) and the values recorded on each of these channels (*Value*). Due to the PBSHM Schema being organic, only rudimentary properties are currently included for each entity.

For the purpose of clarity, this paper defines 'Channel data' as raw sensor data that are directly from sensors. The only processing permitted on this is unit conversion; any additional processing dictates that the data are no longer raw data but

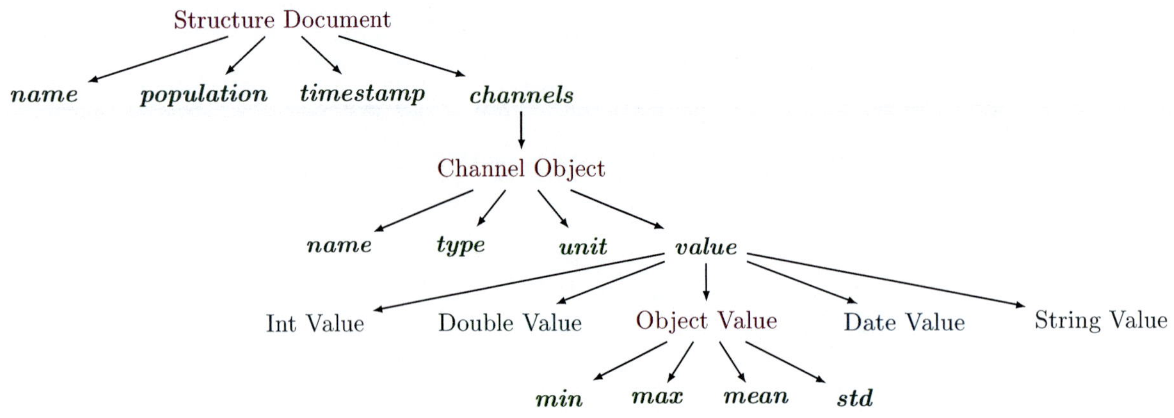

Fig. 7 PBSHM Schema hierarchical structure

Table 1 List of structure object properties in the PBSHM Schema

Property	Description	Type	Required
Name	Name of the structure, must be unique within the population (length greater than 1 character).	*string*	Yes
Population	Name of the population that the structure is part of, must be unique within the PBSHM database (length between 1 and 64 characters).	*string*	Yes
Timestamp	Timestamp of when the associated monitoring data was recorded, stored in UTC nanoseconds since UNIX epoch.	*long*	Yes
Channels	Array of channel objects (see Table 2 & Appendix 3).	*array*	No

instead a feature on the structure and thus should be stored within the yet-to-be-defined 'Feature' section of the PBSHM Schema. For instance, a sensor that operates in units of RPM would be required to be converted into Hz before being permitted into the *Channel* section of the PBSHM Schema due to Hz being the supported base unit.

The *Feature* section of the PBSHM Schema is out of the scope of this paper, however, if a feature is based upon computations of raw sensor data, the raw sensor data preprocessing should be included in the *Channel* section as well as the computed value stored within the *Feature* section, if such preprocessing knowledge is available.

4.3.1 Structure

At the root of the document is knowledge regarding the Time First approach on a Structure: *name*, *population* and *timestamp*. Additional properties can be included within the Structure entity to encapsulate data from within the PBSHM framework; subsequently an additional property has been added for Channel; *channels*. A description of each property within the Structure entity is included in Table 1.

Within MongoDB there is a default *date* property type which stores date information up to millisecond accuracy; however, due to current sensor technology being able to capture data at a 1Mhz sampling rate, this would cause a loss of accuracy on sensor data. Dates within the PBSHM database are implemented via UTC nanoseconds since UNIX epoch and stored in a *long* (Int64) data type which enables sampling up to 1 Ghz.[3]

4.3.2 Channel

Embedded beneath the Structure document is an array of Channel objects containing details on associated Structure sensors; *name*, *type*, *unit* and *value*. A Channel object should exist within the Structure document for each Channel that was present and providing data on the associated Structure at the given Structure timestamp. If a Channel did not provide data at the given timestamp, it should not exist within this Structure document. A description of each property within the Channel entity is included in Table 2.

[3]When using JavaScript to interact with the PBSHM Schema, there will be a loss of accuracy in the timestamps due to JavaScript only supporting 53 bytes in an Int64. This will also affect any interactions within the MongoDB Compass GUI tool as at the time of publishing this paper it is implemented in JavaScript.

Table 2 List of channel object properties in the PBSHM Schema

Property	Description	Type	Required
Name	Name of the channel, must be unique within the structure.	*string*	Yes
Type	The selected type value for this channel, see Appendix 3 for the list of available options.	*string*	Yes
Unit	The selected unit value of this channel which the value is based upon, see Appendix 3 for the list of available options for the selected type.	*string*	See Appendix 3
Value	Value of the channel, stored in the selected unit.	*See Appendix 3*	Yes

Table 3 List of value object properties in the PBSHM Schema

Property	Description	Type
Min	Minimum channel value over the observed time period.	*int, double*
Max	Maximum channel value over the observed time period.	*int, double*
Mean	Mean channel value over the observed time period.	*int, double*
Std	Standard deviation channel value over the observed time period.	*int, double*

To enable accurate comparisons across Channel objects, *type* and *unit* are enumerated types. For each given value of *type* there are associated accepted values for both *unit* and *value*. For instance, if a channel has a value *'tilt'* for *type*, accepted values for *unit* are *'degrees'*, *'radians'* or *'other'* and accepted types for *value* are *int*, *double* or *object*. The *unit* property is not applicable on certain *type*s where having a *unit* to provide context for the *value*s value is not required. For example, a Channel with *type 'text'* does not require any additional context for the value of *value*. A full list of accepted *type*, *unit* and *value* combinations is included in Appendix 3.

4.3.3 Value

When the *unit* for a Channel is applicable, the *value* on the Channel can be of data type *int*, *double* or *object*. A Value object facilitates storage of *value* data which is not singular; *min*, *max*, *mean*, *std*. The Value object must have at least two properties set on the object, otherwise a singular *value* value should be used instead. A description of each property within the Value entity is included in Table 3.

4.3.4 Installation

Setting up a MongoDB Server/Cluster is outside the scope of this paper; however, the latest guidelines from MongoDB can be found in [10, 29]. Once MongoDB is configured in the desired environment, a collection is required with the validation set to the PBSHM Schema. Schema validation can be retrospectively included in a collection; however, for the purposes of this paper it will be assumed that a new collection is created to store PBSHM data and thus have the PBSHM Schema validation. The PBSHM collection is created via the following code using the *pymongo* package within python.

```
db.create_collection(credentials["collection"], validator={
    "$jsonSchema": schema
})
```

A full example detailing how to connect to the server, load the PBSHM Schema into python and create the collection is provided in this paper (see Appendix 4).

4.3.5 Indexes

The PBSHM Schema validates data within the MongoDB collection; however, it does not provide any facility for unique constraints. Indexes can not only improve query performance as outlined, but they also are the mechanism within MongoDB to implement constraints on unique data in a collection. In the example installation script provided (see Appendix 4), the script adds in an index which ensures that every channel name is unique within a single timestamp on a single structure within a single population. For instance, there would not be allowed two channels called 'hello' at the timestamp

```
for document in collection.find(
    {"population":"index-example"},
    {"_id":0}
):
    pp.pprint(document)
```

Fig. 8 Retrieving documents via the *find* method

'1588007839000000000' on the structure 'structure-1' in the population 'population-1'; however, there would be allowed another channel called 'world'. The section of the script that creates this index is provided below.

```
db["collection"].create_index([
    ("population", pymongo.ASCENDING),
    ("name", pymongo.ASCENDING),
    ("timestamp", pymongo.ASCENDING),
    ("channels.name", pymongo.ASCENDING)
], unique=True)
```

The index created in the example script (see Appendix 4) also enables Channel data for a single timestamp on a Structure within a Population to be split across multiple documents (see Sect. 4.3.6 for the cohesion with the PBSHM Schema Requirements). The division of data across multiple documents can provide infrastructure benefits; however, the main advantage to the PBSHM Frameworks is data security and integrity. Because Channel data are split across multiple documents, PBSHM Framework users uploading data only require *insert* and *find* permissions; this means the database can operate as an *immutable* data store.

The advantage to the PBSHM Framework is that data from an end user can be uploaded into the PBSHM database when they become available; this becomes apparent when considering that multiple sub systems may be involved in data capture on a structure. If only a single document is permitted for a timestamp on a Structure within the Population, the end user would create a document when the first set of data was available, then have to find and update the existing document to include additional data. Instead, when permitting multiple documents for a timestamp on a Structure within the Population, the end user is no longer required to find and update existing documents; the end user instead creates a new document for the additional data. An example is provided in this paper which outlines how the same Channel data can be split across multiple files (see Appendix 5).

4.3.6 Querying

The PBSHM Schema Requirements state that retrieval of information from the PBSHM database must be easy to understand; however, due to data Channel data being stored as either a single document or multiple documents for a given timestamp on a Structure within a Population, using the standard MongoDB *find* method (see Fig. 8) for querying the database would produce different results depending on the setup of the environment.

MongoDB provides an *aggregate* method which enables processing data before returning results. Using the *aggregate* method, results can be standardised regardless of the single/multiple document environment. The *aggregate* method takes an array of *aggregate operations* (*pipeline*) to perform the data processing. The *match* operation of an *aggregate* method is identical to the *query* parameter of the *find* method.

The *aggregate* example included (see Fig. 9) performs the following operations; *unwind*s the Structure documents into virtual Structure document, creating a document for every Channel within a Structure document such that each virtual Structure document contains only one nested Channel object, *group*s all of the virtual documents based upon the *population*, *name* and *timestamp* values and then removes the *_id* created to group together the documents on the returned document.

A full example detailing how to connect to the server and query the PBSHM Database via both the *find* and *aggregate* methods is provided in this paper (see Appendix 6).

```
for document in collection.aggregate([
    {"$match": {"population": "index-example"}},
    {"$unwind": {"path": "$channels"}},
    {
        "$group": {
            "_id": {
                "population": "$population",
                "name": "$name",
                "timestamp": "$timestamp"
            },
            "name": {"$first": "$name"},
            "population": {"$first": "$population"},
            "timestamp": {"$first": "$timestamp"},
            "channels": {"$addToSet": "$channels"}
        }
    },
    {"$project": {"_id": 0}}
]):
    pp.pprint(document)
```

Fig. 9 Retrieving documents via the *aggregate* method

Sensor Group 1 CSV Data File

01-01-2020:00:00:00	Structure1	9.5	10.0	9.75	0	starting
01-01-2020:00:00:00	Structure2	8.5	9.0	8.75	0	about to start
01-01-2020:00:00:00	Structure3	7.5	8.0	7.75	0	finished
01-01-2020:00:00:00	Structure4	6.5	7.0	6.75	0	starting
01-01-2020:00:00:00	Structure5	5.5	6.0	5.75	0	finished

(a)

Sensor Group 1 Header File

Time	Name	Sensor1_min	Sensor1_max	Sensor1_mean	Sensor1_stddev	Sensor2_info

(b)

Fig. 10 Example of Group 1 set of raw data on the Vattenfall Lillgrund data set. (**a**) An example of raw Channel data for Group 1 set of sensors. (**b**) An example of header data for Group 1 set of sensors

4.3.7 Download

The PBSHM Schema included in this paper can be downloaded from https://github.com/dynamics-research-group/pbshm-schema/releases/tag/v1.0.1.[4] A full list of supported enumerated types and further documentation is available from the Dynamics Research Groups PBSHM Schema GitHub Repository https://github.com/dynamics-research-group/pbshm-schema.

4.4 Data Import

To ensure that the PBSHM Schema can hold real world data, the Vattenfall Lillgrund data set has been imported into the PBSHM database for illustration here. The Lillgrund data set is 10-min SCADA data for a population of 48 offshore wind turbines. Channel data is separated into six files, three CSV files which contain the SCADA data and three header files which contain the name of the column at the same index in their associated data file (see Fig. 10). The data set has missing values as well as duplicate values for certain sections of the data set.

[4]The file for installation within MongoDB is *structure-data-compiled-mongodb.min.json*.

```
{
    "name": "Structure1",
    "population": "vattenfall-lillgrund",
    "timestamp": 1577836800000000000,
    "channels": [
        {
            "name": "Sensor1",
            "type": "temperature",
            "unit": "C",
            "value": {
                "min": 9.5,
                "max": 10.0,
                "mean": 9.75,
                "std": 0
            }
        },
        {
            "name": "Sensor2",
            "type": "text",
            "value": "starting"
        }
    ]
}
```

Fig. 11 Example of Group 1 set of data on the Vattenfall Lillgrund data set stored in the PBSHM database

An importer has been developed to read all of the header document within the data set and create a sensor mapping dictionary based upon the underscore-separated columns names. For example, the importer associates *Sensor1_min*, *Sensor1_max*, *Sensor1_mean* and *Sensor1_stddev* as all being *Sensor1* with the properties being *min*, *max*, *mean* and *stddev* at the indexes of *2*, *3*, *4* and *5*. An additional file is also passed to the importer adding in missing sensor mapping information that the PBSHM Schema requires; *type* and *unit*.

Due to the PBSHM Schema separating Structure documents based upon the *name* and *timestamp*, the importer can use multiple threads to process each line of the data file as each line represents a timestamp on a structure. The importer has been implemented via C#, thus instead of storing the files as JSON within the program before insertion into MongoDB, it uses the native BSON data objects available in the MongoDB C# driver. An example of the transformed data is included in this paper (see Fig. 11).

For example, once the data are within the PBSHM Schema format, the database can be queried for 'temperature' data within the 'vattenfall-lillgrund' population:

```
for document in collection.find(
    {"population": "vattenfall-lillgrund", "channels.type": "temperature"},
    {
        "_id": 0, "name": 1, "population": 1, "timestamp": 1,
        "channels": {"$elemMatch": {"type": "temperature"}}
    }
):
    pp.pprint(document)
```

The source code for the importer included in this paper can be downloaded from https://github.com/dynamics-research-group/drg-dbimport/releases/tag/v1.0. For more documentation and the latest importer examples, please visit the Dynamics Research Group DRG Import GitHub Repository https://github.com/dynamics-research-group/drg-dbimport.

5 Conclusion

This paper has discussed the requirements for a data schema for PBSHM and discussed alternative implementations of this. Suggestions have been made for how we believe is the best approach to solve this issue and examples have been provided for both querying the database and populating the database with historic data. Although this paper has suggested a NoSQL Document database approach, an RDBMS approach would potentially facilitate improved read performance for particular queries with a smaller data set where channel data is known before the development of the database and would be consistent throughout the lifetime of the system. NoSQL Graph databases may also provide a more natural approach for a PBSHM database where data within the database would be only of a node/edge type such as a graph matching scenario.

The suggested PBSHM Schema provides a unified approach to the storage and retrieval of PBSHM data, thus enabling a user-friendly means in which to compare multiple structures across different populations without the prerequisite of expert knowledge on each individual data set. The next steps are to expand the knowledge of the PBSHM Schema to include feature data and attributed graphs. Once this is complete, the core technology for the PBSHM Framework can be developed and subsequent development on PBSHM Modules can begin.

Acknowledgments The authors of this paper gratefully acknowledge the support of the UK Engineering and Physical Sciences Research Council (EPSRC) through grant references EP/J016942/1, EP/K003836/2 and EP/S001565/1. The authors would also like to thank Vattenfall and Eoghan Maguire for providing the data set used within this paper.

Appendix 1: Non-organic Basic Schema

Please note that the example below uses the MongoDB bsonType instead of the JSON

```
{
    "bsonType": "object",
    "required": [
        "name", "channel1Name", "channel1Data", "channel2Name", "channel2Data"
    ],
    "properties": {
        "name": {
            "bsonType": "string",
            "title": "Structure Name"
        },
        "channel1Name": {
            "bsonType": "string",
            "title": "Channel 1 Name"
        },
        "channel1Data": {
            "bsonType": "array",
            "title": "Channel 1 Data"
        },
        "channel2Name": {
            "bsonType": "string",
            "title": "Channel 2 Name"
        },
        "channel2Data": {
            "bsonType": "array",
            "title": "Channel 2 Data"
        }
    }
}
```

Appendix 2: Organic Basic Schema

Please note that the example below uses the MongoDB bsonType instead of the JSON

```
{
    "bsonType": "object",
    "required": [
        "name", "channels"
    ],
    "properties": {
        "name": {
            "bsonType": "string",
            "title": "Structure Name"
        },
        "channels": {
            "bsonType": "array",
            "title": "Channel Array",
            "items": {
                "bsonType": "object",
                "required": [
                    "name", "data"
                ],
                "properties": {
                    "name": {
                        "bsonType": "string",
                        "title": "Channel Name"
                    },
                    "data": {
                        "bsonType": "array",
                        "title": "Channel Data"
                    }
                }
            }
        }
    }
}
```

Appendix 3: Accepted Channel Objects Within the v1.0 Schema

Types	Units	Values
acceleration	*m/s^2, g, v, other*	*int, double, object*
velocity	*m/s, v, other*	*int, double, object*
displacement	*mm, cm, m, km, other*	*int, double, object*
angularAcceleration	*degrees/s^2, radians/s^2, other*	*int, double, object*
angularVelocity	*degrees/s, radians/s, other*	*int, double, object*
angularDisplacement	*degrees, radians, other*	*int, double, object*
tilt	*degrees, radians, other*	*int, double, object*
strain	*nd, other*	*int, double, object*
tension	*fN, pN, nN, μN, mN, cN, dN, N, daN, hN, kN, MN, GN, TN, PN, other*	*int, double, object*
load	*fN, pN, nN, μN, mN, cN, dN, N, daN, hN, kN, MN, GN, TN, PN, other*	*int, double, object*
structuralPotentialHydrogen	*pH, other*	*int, double, object*
temperature	*C, F, K, other*	*int, double, object*
humidity	*%, other*	*int, double, object*
speed	*mph, ft/s, km/h, m/s, kn, other*	*int, double, object*
direction	*degrees, radians, other*	*int, double, object*
pressure	*fPa, pPa, nPa, μPa, mPa, cPa, dPa, Pa, daPa, hPa, kPa, MPa, GPa, TPa, PPa, at, atm, bar, psi, other*	*int, double, object*
altitude	*mm, cm, m, km, feet, other*	*int, double, object*
pitch	*degrees, radians, other*	*int, double, object*
yaw	*degrees, radians, other*	*int, double, object*
roll	*degrees, radians, other*	*int, double, object*
pitchRate	*degrees/s, radians/s, other*	*int, double, object*
yawRate	*degrees/s, radians/s, other*	*int, double, object*
rollRate	*degrees/s, radians/s, other*	*int, double, object*
current	*fA, pA, nA, μA, mA, cA, dA, A, daA, hA, kA, MA, GA, TA, PA, other*	*int, double, object*
charge	*fC, pC, nC, μC, mC, cC, dC, C, daC, hC, kC, MC, GC, TC, PC, other*	*int, double, object*
power	*fW, pW, nW, μW, mW, cW, dW, W, daW, hW, kW, MW, GW, TW, PW, other*	*int, double, object*
voltage	*fV, pV, nV, μV, mV, cV, dV, V, daV, hV, kV, MV, GV, TV, PV, other*	*int, double, object*
resistance	*fΩ, pΩ, nΩ, μΩ, mΩ, cΩ, dΩ, Ω, daΩ, hΩ, kΩ, MΩ, GΩ, TΩ, PΩ, other*	*int, double, object*
capacitance	*fF, pF, nF, μF, mF, cF, dF, F, daF, hF, kF, MF, GF, TF, PF, other*	*int, double, object*
inductance	*fH, pH, nH, μH, mH, cH, dH, H, daH, hH, kH, MH, GH, TH, PH, other*	*int, double, object*
frequency	*fHz, pHz, nHz, μHz, mHz, cHz, dHz, Hz, daHz, hHz, kHz, MHz, GHz, THz, PHz, other*	*int, double, object*
conductance	*fS, pS, nS, μS, mS, cS, dS, S, daS, hS, kS, MS, GS, TS, PS, other*	*int, double, object*
magneticFlux	*fWb, pWb, nWb, μWb, mWb, cWb, dWb, Wb, daWb, hWb, kWb, MWb, GWb, TWb, PWb, other*	*int, double, object*
magneticFieldStrength	*fT, pT, nT, μT, mT, cT, dT, T, daT, hT, kT, MT, GT, TT, PT, other*	*int, double, object*
integer	n/a	*int*
double	n/a	*double*
text	n/a	*string*
date	n/a	*long*

Appendix 4: PBSHM Schema Installation

An example python script for installing the PBSHM Schema into a collection.

```
import json
import pymongo
import urllib.parse

# Credentials
credentials = {
    "host": "localhost", "port": "27017",
    "username": "", "password": "", "authdb": "admin",
    "database": "", "collection": ""
}

# Connect to Server
client = pymongo.MongoClient("mongodb://{username}:{password}@{host}:{port}/{authdb}".format(
    username=urllib.parse.quote_plus(credentials["username"]),
    password=urllib.parse.quote_plus(credentials["password"]),
    host=credentials["host"], port=credentials["port"], authdb=credentials["authdb"]
))

# Load Schema
with open("./release/structure-data-compiled-mongodb.min.json") as file:
    schema = json.load(file)

# Create Collection
db = client[credentials["database"]]
db.create_collection(credentials["collection"], validator={
    "$jsonSchema": schema
})

# Create Unique Index
db[credentials["collection"]].create_index([
    ("population", pymongo.ASCENDING),
    ("name", pymongo.ASCENDING),
    ("timestamp", pymongo.ASCENDING),
    ("channels.name", pymongo.ASCENDING)
], unique=True)
```

Appendix 5: Single vs Multiple File Channel Data

A JSON example showing equal Channel Data on a single file vs multiple files.


```
{
    "name": "structure-one",
    "population": "index-example",
    "timestamp": 1588007839000000000,
    "channels": [
        {
            "name": "channel-1",
            "type": "acceleration",
            "unit": "g",
            "value": {
                "min": 9.5,
                "max": 10.5,
                "mean": 10,
                "std": 0.5
            }
        },
        {
            "name": "channel-2",
            "type": "temperature",
            "unit": "C",
            "value": 11.5
        },
        {
            "name": "channel-3",
            "type": "text",
            "value": "text based context"
        }
    ]
}
```


(a) Channel data stored in a single file.


```
{
    "name": "structure-one",
    "population": "index-example",
    "timestamp": 1588007839000000000,
    "channels": [
        {
            "name": "channel-1",
            "type": "acceleration",
            "unit": "g",
            "value": {
                "min": 9.5,
                "max": 10.5,
                "mean": 10,
                "std": 0.5
            }
        }
    ]
}
```


(b) Channel 1 data stored in it's own file.


```
{
    "name": "structure-one",
    "population": "index-example",
    "timestamp": 1588007839000000000,
    "channels": [
        {
            "name": "channel-2",
            "type": "temperature",
            "unit": "C",
            "value": 11.5
        },
        {
            "name": "channel-3",
            "type": "text",
            "value": "text based context"
        }
    ]
}
```


(c) Channel 2 & 3 data stored in a second file.

Appendix 6: PBSHM Database Querying

An example python script for querying the PBSHM Database.

```python
import json
import pymongo
import urllib.parse
import pprint

# Credentials
credentials = {
    "host": "localhost", "port": "27017",
    "username": "", "password": "", "authdb": "admin",
    "database": "", "collection": ""
}
# Connect to Server
client = pymongo.MongoClient("mongodb://{username}:{password}@{host}:{port}/{authdb}".format(
    username=urllib.parse.quote_plus(credentials["username"]),
    password=urllib.parse.quote_plus(credentials["password"]),
    host=credentials["host"], port=credentials["port"], authdb=credentials["authdb"]
))

# Setup Variables Document
pp = pprint.PrettyPrinter(indent=4)
db = client[credentials["database"]]
collection = db[credentials["collection"]]

#Standard Retrieve
for document in collection.find(
    {"population":"index-example"},
    {"_id":0}
):
    pp.pprint(document)

#Aggregate Retrieve
for document in collection.aggregate([
    {"$match": {"population": "index-example"}},
    {"$unwind": {"path": "$channels"}},
    {
        "$group": {
            "_id": {
                "population": "$population",
                "name": "$name",
                "timestamp": "$timestamp"
            },
            "name": {"$first": "$name"},
            "population": {"$first": "$population"},
            "timestamp": {"$first": "$timestamp"},
            "channels": {"$addToSet": "$channels"}
        }
    },
    {"$project": {"_id": 0}}
]):
    pp.pprint(document)
```

References

1. Worden, K., Cross, E.J., Dervilis, N., Papatheou, E., Antoniadou, I.: Structural health monitoring: from structures to systems-of-systems. Int. Fed. Autom. Control **48**(21), 1–17 (2015)
2. Bull, L.A., Gardner, P.A., Gosliga, J., Rogers, T.J., Dervilis, N., Cross, E.J., Papatheou, E., Maguire, A.E., Campos, C., Worden, K.: Foundations of population-based SHM, Part I: homogeneous populations and forms. Mech. Syst. Signal Process. **148**, 107141 (2020)
3. Worden, K., Bull, L.A., Gardner, P.A., Gosliga, J., Rogers, T.J., Cross, E.J., Papatheou, E., Lin, W., Dervilis, N.: A brief introduction to recent developments in population-based structural health monitoring. Frontiers (2020). https://doi.org/10.3389/fbuil.2020.00146
4. Jeong, S., Byun, J., Kim, D., Sohn, H., Bae, H., Law, K.H.: A data management infrastructure for bridge monitoring. In: SPIE - The International Society for Optical Engineering (2015)
5. Wickramarachchi, C., Brennan, D.S., Weijiang, L., Harvey, D., Cross, E.J., Worden, K.: Towards a Population-Based Structural Health Monitoring: Part V, networks and databases. In: IMAC XXXIX (2021)
6. Codd, E.F.: A relational model of data for large shared data banks. **13**(6), 377–387 (1970)
7. Nance, C., Losser, T., Iype, R., Harmon, G.: NoSQL vs RDBMS - Why there is room for both. In: SAIS 2013 Proceedings, vol. 27 (2013)
8. Zafar, R., Yafi, E., Zuhairi, M.F., Dao, H.: Big Data: The NoSQL and RDBMS review. In: International Conference on Information and Communication Technology (2016)
9. Gandini, A., Gribaudo, M., Knottenbelt, W.J., Osman, R., Piazzolla, P.: Performance evaluation of NoSQL databases. In: European Workshop on Performance Engineering (2014)
10. Bradshaw, S., Brazil, E., Chodorow, K.: MongoDB: The Definitive Guide, 3rd edn. O'Reilly, Sebastopol (2019)
11. MongoDB, Inc.: Encryption at Rest - MongoDB Manual. Accessed 6 April 2020
12. MongoDB, Inc.: Users - MongoDB Manual. Accessed 6 April 2020
13. MongoDB, Inc.: TLS/SSL (Transport Encryption) - MongoDB Manual. Accessed 6 April 2020
14. Codd, E.F.: Further normalization of the data base relational model. In: Rustin, R. (ed.) Data Base Systems (Courant Computer Science Symposia 6). Prentice-Hall, Englewood Cliffs (1972)
15. Crawford, D.: Technical correspondence. Commun. ACM **36**(10), 19 (1993)
16. MongoDB, Inc. About Us - MongoDB. Accessed 9 April 2020
17. Bray,T.: RFC 7159 - The JavaScript Object Notation (JSON) Data Interchange Format. Accessed 15 April 2020
18. Bray, T.: RFC 8259 - The JavaScript Object Notation (JSON) Data Interchange Format. Accessed 15 April 2020
19. MongoDB, Inc.: Role-Based Access Control - MongoDB Manual. Accessed 6 April 2020
20. MongoDB, Inc.: Storage Engines - MongoDB Manual. Accessed 6 April 2020
21. MongoDB, Inc.: WiredTiger Storage Engine - MongoDB Manual. Accessed 6 April 2020
22. MongoDB, Inc.: Indexes - MongoDB Manual. Accessed 15 April 2020
23. MongoDB, Inc.: Replication - MongoDB Manual. Accessed 15 April 2020
24. MongoDB, Inc.: Sharding - MongoDB Manual, Accessed 15th April, 2020.
25. MongoDB, Inc.: Data Modeling Introduction - MongoDB Manual. Accessed 6 April 2020
26. MongoDB, Inc.: Normalized Data Models - MongoDB Manual. Accessed 16 April 2020
27. MongoDB, Inc.: Embedded Data Models - MongoDB Manual. Accessed 16 April 2020
28. MongoDB, Inc.: MongoDB Limits and Thresholds - MongoDB Manual. Accessed 6 April 2020
29. MongoDB, Inc.: Install MongoDB - MongoDB Manual. Accessed 18 May 2020

Estimation of Blade Forces in Wind Turbines Using Strain Measurements Collected on Blades

Bridget Moynihan, Babak Moaveni, Sauro Liberatore, and Eric Hines

Abstract This paper introduces a method for computing the forces and bending moments on operating wind turbine blades based on strain and SCADA data in the absence of comprehensive design specifications. Operational data from 4 months of a Clipper Liberty C96 2.5 MW turbine instrumented with interferometric strain sensors placed at the blade roots as well as supervisory Control and Data Acquisition (SCADA) data such as wind speed, rotor hub speed, and blade pitch angle allow for accurate calculation of blade forces and moments. In order to perform such calculations, certain structural properties of the turbine blades must be established in the absence of detailed, proprietary information. This is done by inferring certain properties from National Renewable Energy Laboratory (NREL) reference wind turbine specifications. In these specifications, NREL also provides descriptions of resulting forces and other parameters to be used in the design of wind turbines. The authors independently computed these forces and moments as a function of wind speed and compared them to NREL's results in the paper, demonstrating that this method can be used to successfully estimate the internal forces and bending moments acting on the blades.

Keywords Wind turbine · Condition monitoring · Strain sensors · Wind energy · Structural health monitoring

1 Introduction

In 2019 alone, 39% of new additions to the U.S. electricity supply came from wind power, and total U.S. wind power capacity has now surpassed 100 GW, which is four times the existing operating capacity from 10 years ago [1]. As the wind industry has grown, so have the wind turbines (WT) themselves. As OWTs grow in size, the reliability of such massive structures is a key consideration to ensure the availability of low-cost renewable energy. One of the largest components of wind's levelized cost of energy (LCOE) is that of operation and maintenance (O&M), which can constitute 10–20% of the LCOE in WT and reach up to 35% at the end of life [2]. O&M not only increases total costs, but it creates downtime which can reach up to 3% of WT and OWT total lifetimes [3].

Structural health monitoring (SHM) as a form of condition monitoring (CM) is an important tool for ensuring the integrity of WT infrastructure [4], as it allows for maintenance before failure occurs, reducing O&M costs [5]. Application of SHM methods on strain measurement is well studied but obtaining further information on the condition of WT from strain measurements can make CM systems more cost efficient. In traditional CM systems, strain sensors are typically placed on the gearbox, generators, and drivetrain [6] more commonly than on other major components such as the blades or tower. Some WT are instrumented with strain sensors on additional components such as the blades. Fiber optic strain gauges placed on blades can be used for CM of blades, though it is one of the most expensive approaches compared to other methods [7]. This type of CM is utilized by WT operators to measure blade-root bending moments [7].

Analyzing the root reaction forces and bending moments on the blades in addition to blade-root bending moments would be useful for understanding the effects of certain loading conditions [8], but to utilize strain measurements for making these calculations, WT design details must be known.

B. Moynihan (✉) · B. Moaveni · E. Hines
Tufts University, Medford, MA, USA
e-mail: bridget.moynihan@tufts.edu; babak.moaveni@tufts.edu; eric.hines@tufts.edu

S. Liberatore
MIDE Technology Corporation, Woburn, MA, USA
e-mail: sliberatore@mide.com

K. Grimmelsman (ed.), *Dynamics of Civil Structures, Volume 2*, Conference Proceedings of the Society for Experimental Mechanics Series, https://doi.org/10.1007/978-3-030-77143-0_4

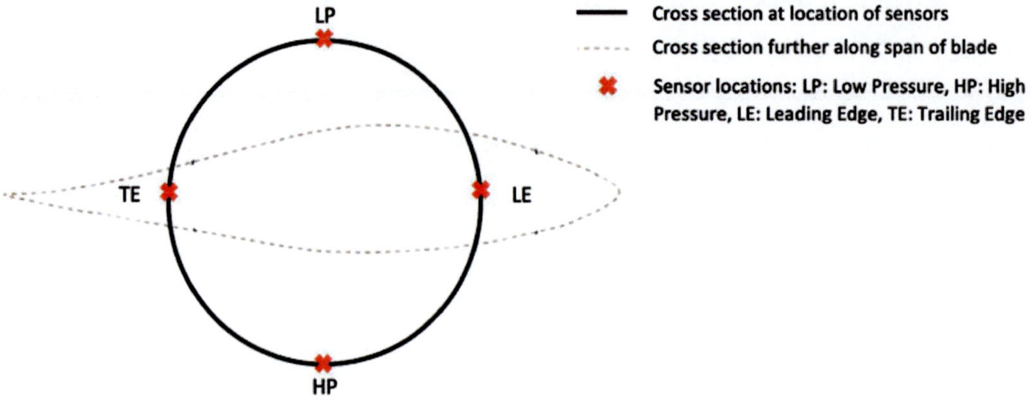

Fig. 1 Sensor location at blade roots

This paper details a method to analyze the forces on the blades of a Clipper Liberty C96 2.5 MW turbine instrumented with strain sensors at the blade roots. SCADA data from the turbine is also available and used in this analysis. This study computes the forces and bending moments on the blades from the strain measurements in the absence of all the detailed design specifications. The reference WT detailed by the National Renewable Energy Laboratory (NREL) is used to infer certain structural properties of the WT blades where proprietary information from manufacturers is unavailable [9]. In these specifications, NREL also provides descriptions of resulting forces and other parameters to be used in the design of WT. Results of this method are compared to these forces and parameters described by NREL as a function of wind speed.

2 Methodology

The data used in this paper includes axial strain measurements from the roots of each of the blades, where four sensors are attached. This data is collected in hour long periods spanning 4 months of operation and over 1500 h of data is available. Supervisory Control and Data Acquisition (SCADA) data is available from the same time period and the data are synchronized with the sensor readings. The strain sensors used are optical intereferometers with corresponding temperature sensors as a way to compensate for temperature. The sensors are positioned in the four locations at each blade root schematically depicted in Fig. 1.

A rigid model for the considered section wind turbine was used due to its high stiffness [10] and displayed in Fig. 2. Moments and forces on each of the blades are calculated in the local axes of the blade, which can be resolved onto the rotor in the global axes system.

The local axes of blades are labeled as x, y, z, while the global axes of turbine are denoted as X, Y, Z as shown in Fig. 2. The x-axis is perpendicular to span of blade, and points in the same direction as the nacelle, running parallel to and coincident with wind direction (when there is no veer). The y-axis is in the axial direction and along the blade, and the z-axis is in the plane of the blades' rotation.

Two key bending moments in flapwise and edgewise directions can be calculated directly from the strain data due to the placement of the sensors at the High Pressure, Low Pressure, Leading Edge, and Trailing Edge positions. The flapwise and edgewise moments are calculated from the temperature-compensated strain data as described by Eqs. (2) and (3). These moments are later resolved into the local blade axes defined for our analysis.

$$M_{\text{Flapwise}} = (\varepsilon_{\text{HP}} - \varepsilon_{\text{LP}}) * k1 \tag{1}$$

$$M_{\text{Edgewise}} = (\varepsilon_{\text{LE}} - \varepsilon_{\text{TE}}) * k2 \tag{2}$$

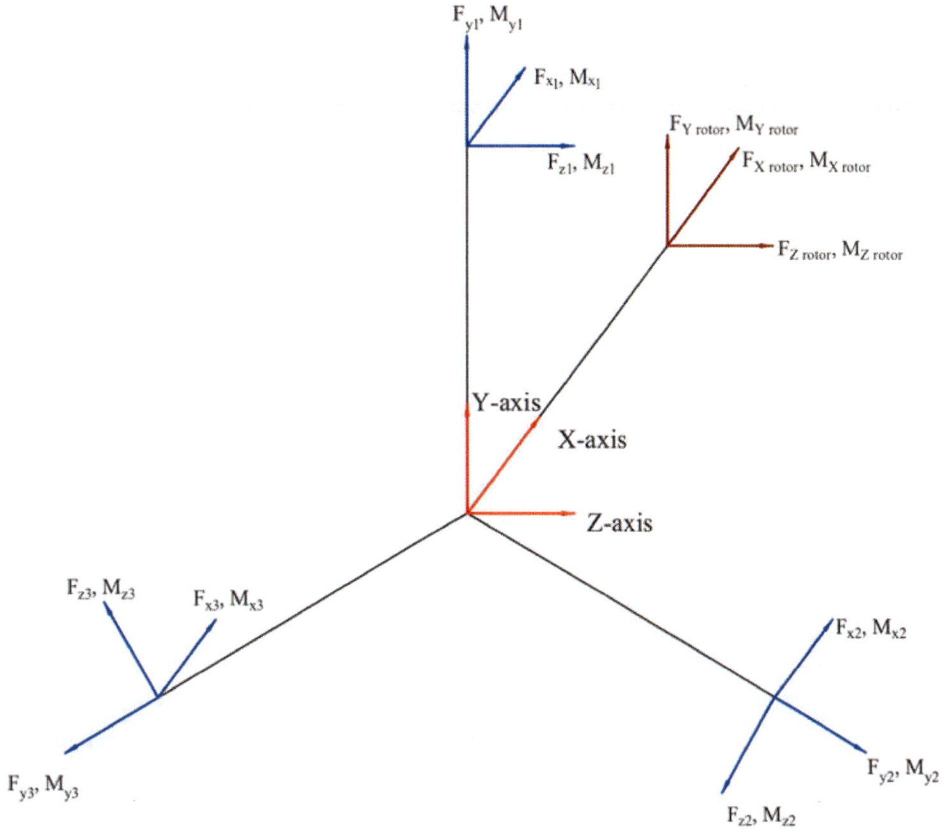

Fig. 2 Local and global axes coordinate system

In above equations, $k1$ and $k2$ are constants provided by the sensor calibration data, and ε_{HP}, ε_{LP}, ε_{TE}, and ε_{LE} are the temperature-compensated strain measurements.

The sensor data from the blades cannot provide direct calculations of shear forces because the sensors are placed along the axial direction. These shear forces in local coordinates are in the x–z plane of the blades. These shear forces, F_x and F_z, coupled with the effect of blade weight, would cause the flapwise and edgewise bending moments. This means that since the flapwise and edgewise moments and the weight of the blades are known, the magnitude of the shear forces can be calculated. This is done by defining the flapwise and edgewise bending moments as moments caused by lift and drag forces on the blades plus the effect of blade weight as the blades rotate, as described in Eqs. (3) and (4). Effective lift and effective drag forces, $F_{\text{drag,eff}}$ and $F_{\text{lift,eff}}$, are defined as forces which act in the flapwise and edgewise directions in the x–z plane. The effective lift and drag forces will be resolved into our axes once their magnitudes are known. The effective lift and drag forces are defined as the resultant of distributed lift and drag forces along the blades. These forces are computed using Eqs. (3) and (4), as $F_{\text{drag,eff}}$ and $F_{\text{lift,eff}}$ are the only unknowns here.

$$M_{\text{Flap}} = (\text{mg})_{z,\text{flap}} d_{\text{CG}} + F_{\text{lift,eff}} d_{\text{L}} \tag{3}$$

$$M_{\text{Edge}} = -(\text{mg})_{z,\text{edge}} d_{\text{CG}} - F_{\text{drag,eff}} d_{\text{D}} \tag{4}$$

In the above equations, $(\text{mg})_{z,\text{ flap}}$ and $(\text{mg})_{z,\text{ edge}}$ are the components of blade weight laying in the local x–z axis of the blade and resolved into the flapwise and edgewise directions. The distance d_{CG} is the center of gravity of the blades, and d_{L} and d_{D} are the distances at which resultant lift and drag forces act along the span of the blade. The distances d_{L} and d_{D} have been computed based on the detailed blade profile of the NREL 3 MW reference turbine, as the blade properties of the Clipper 2.5 MW turbine are not known.

Once $F_{\text{lift,eff}}$ and $F_{\text{drag,eff}}$ are computed, they can be resolved into the local defined axes of the blades to define the shear forces on the blades in the x and z directions. This is based on the pitch angle, β, and is defined by Eqs. (5) and (6). Similarly, the flapwise and edgewise moments which act in the same directions as $F_{\text{lift,eff}}$ and $F_{\text{drag,eff}}$ can be resolved to compute the moments on the blades in the x and z directions according to Eqs. (7) and (8).

$$F_x = F_{\text{drag,eff}} \sin(\beta) + F_{\text{lift,eff}} \cos(\beta) \tag{5}$$

$$F_z = -F_{\text{drag,eff}} \cos(\beta) + F_{\text{lift,eff}} \sin(\beta) \tag{6}$$

$$M_x = M_{\text{flapwise}} \sin(\beta) + M_{\text{Edgewise}} \cos(\beta) \tag{7}$$

$$M_z = -M_{\text{flapwise}} \cos(\beta) + M_{\text{Edgewise}} \sin(\beta) \tag{8}$$

The axial force in each blade is affected by the wind conditions and the centrifugal forces due to blade rotation. Since the strain measured at the blade root are available, their value can be obtained directly from the strain sensors. The axial force is computed from Eq. (9):

$$F_n = \frac{(\varepsilon_{\text{HP}} + \varepsilon_{\text{LP}} + \varepsilon_{\text{TE}} + \varepsilon_{\text{LE}})}{4} * E * A \tag{9}$$

where ε_{HP}, ε_{LP}, ε_{TE}, and ε_{LE} are the temperature-compensated strain measurements at the blade root. Young's modulus, E, and cross-sectional area, A, are unknown and must be inferred from NREL reference turbines. The moment in the axial direction of each blade is taken as the torque which is provided by the SCADA data.

3 Results

The NREL reference wind turbine report produces a graph of certain forces and other parameters of the turbine during operation as a function of wind speed. These include parameters such as rotor speed, pitch angle, and tip speed ratio (TSR), and the rotor thrust force and rotor torque. The graphs produced by NREL are shown in Fig. 3. Figure 4 graphs the parameters and forces computed from the strain sensors and SCADA data in this analysis for comparison with NREL reference turbine values.

Rotor thrust for the Clipper 2.5 MW turbine is computed through a rigid body equilibrium analysis by resolving the forces from each of the blades onto the main shaft as a function of blade azimuth angle. The comparison of rotor thrust values shows that while our values are slightly lower in magnitude, the shape of the force as a function of wind speed is similar. In both cases, rotor thrust rises until approximately 11 m/s, the rated wind speed of the NREL 3 MW reference turbine, before declining again. This aligns well with the expected rotor thrust shape for the Clipper turbine, as the rated wind speed for the Clipper 2.5 MW turbine is 12.5 m/s. The slight differences in shape and lower magnitude of rotor thrust could be explained by the different rated wind speeds, and the fact that the Clipper turbine is smaller in both size and output. In fact, the rotor thrust magnitudes computed for the Clipper 2.5 MW turbine here fall somewhere between rotor thrust values for the NREL 1.5 MW and 3.0 MW reference turbines.

Other parameters such as rotor speed, pitch angle, and tip speed ratio (TSR) compared here come from the SCADA data directly or are calculated using parameters in the SCADA data. These parameters match the reference turbine well, as they are products of the control system on the turbine and independent of the calculations in this paper. These parameters, in particular the rotor speed, match the reference turbine closely, which serves as confirmation that using the 3 MW turbine as reference for this Clipper 2.5 MW is a reasonable estimation to make.

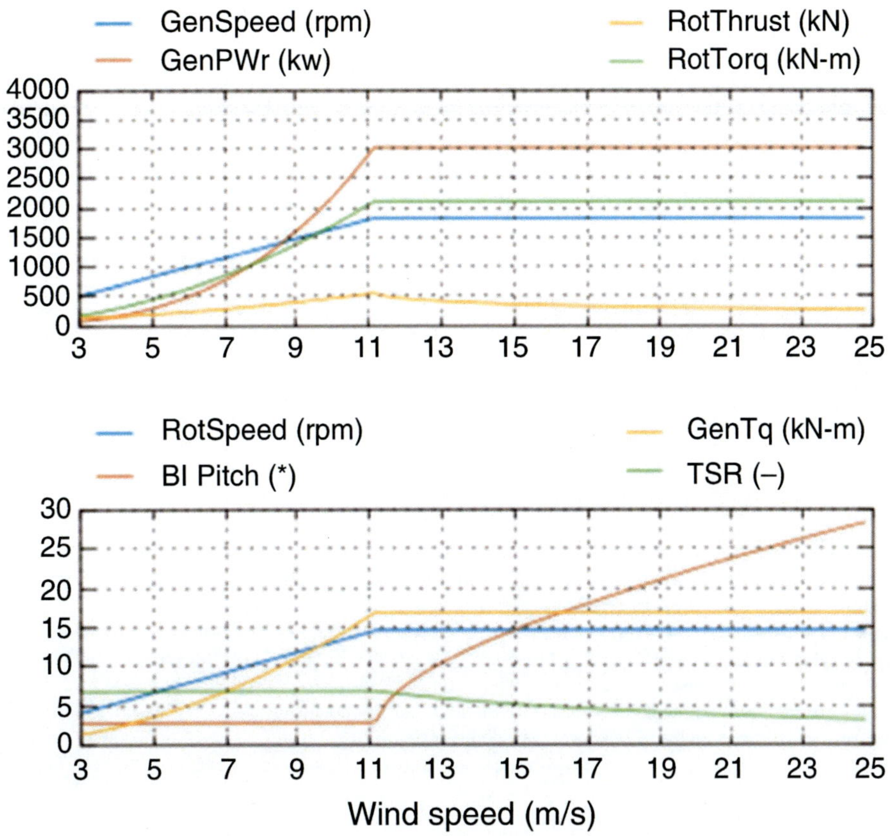

Fig. 3 NREL 3 MW reference turbine parameters

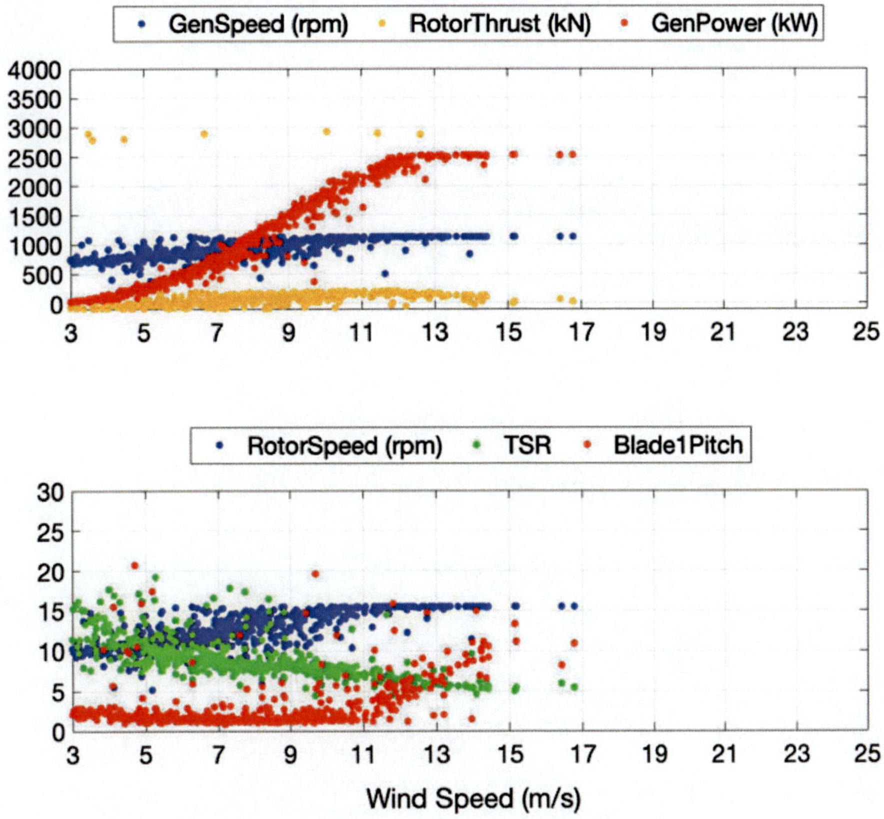

Fig. 4 Parameters Computed based on Strain data and SCADA data for the Clipper 2.5 MW Turbine

4 Conclusions

The methods detailed in this paper allow for the computation of blade-root reaction forces and moments based on strain measurements taken at the roots by using parameters of an NREL reference turbine of a similar size and output to replace structural design details not available to the public. Initial results show that this method is a viable way to estimate the blades' forces and moments, as the resolution of these forces onto the rotor results in a close match with the rotor forces according to the NREL reference turbine parameters. Discrepancies in this comparison may be due to the slight difference in size and different output capacities of the Clipper 2.5 MW turbine considered in this paper and the NREL 3 MW WindPACT reference turbine.

Acknowledgments Partial support of the study by the National Science Foundation grant 1903972 is gratefully acknowledged.

References

1. Wind Powers America Annual Report. Executive Summary, American Wind Energy Association. https://www.awea.org/getattachment/Resources/Publications-and-Reports/Market-Reports/AMR2019_ExecutiveSummary/AMR2019_ExecutiveSummary_Download/AWEA_WPA_ExecutiveSummary2019.pdf.aspx?lang=en-US (2019)
2. Operation and Maintenance Costs. Sandia Report No. SAND2006-1100. Sandia National Laboratories, Albuquerque, NM (2006)
3. Tchakoua, P., Wamkeue, R., Ouhrouche, M., Hasnaoui, F., Tameghe, T., Andy, T., Ekemb, G.: Wind turbine condition monitoring: state-of-the-art review, new trends, and future challenges. Energies. 7, 2595–2630 (2014). https://doi.org/10.3390/en7042595
4. Ribrant, J.: Reliability performance and maintenance—A survey of failures in wind power systems. Master's Thesis, School of Electrical Engineering, KTH Royal Institute of Technology, Stockholm, Sweden (2005–2006)
5. Pedregal, D.J., Garcia Marquez, F.P., Roberts, C.: An algorithmic approach for maintenance management. Ann. Oper. Res. **166**, 109–124 (2009)
6. Condition monitoring. Wind Power Monthly. https://www.windpowermonthly.com/condition-monitoring-systems
7. Yang, W., Tavner, P.J., Crabtree, C., Feng, Y., Qiu, Y.: Wind turbine condition monitoring: technical and commercial challenges. Wind Energy. **17** (2014). https://doi.org/10.1002/we.1508
8. Li, Y., et al.: Research and development of the wind turbine reliability. Int. J. Mech. Eng. Appl. **6**(2) (2018)
9. Rinker, J., Dykes, K.: WindPACT Reference Wind Turbines. National Renewable Energy Laboratory Technical Report NREL/TP-5000-67667. www.nrel.gov/publications (2018)
10. Fava, G., et al.: Estimating fatigue in the main bearings of wind turbines using experimental data. In: Conference Proceedings of the Society for Experimental Mechanics Series (2018)

On Health-State Transition Models for Risk-Based Structural Health Monitoring

A. J. Hughes, R. J. Barthorpe, and K. Worden

Abstract A desire for informed decision-making regarding the operation and maintenance of structures provides motivation for the development and implementation of structural health monitoring (SHM) systems. One approach to decision-making in SHM is to adopt a risk-based framework in which failure events and decidable actions are attributed costs/utilities. Optimal maintenance strategies may be pursued by considering the probability of occurrence of future failure events in conjunction with associated costs. In order to forecast future failure events, a probabilistic model that describes the degradation of the structure over time is required; in the state-space formulation of risk-based SHM, this model is equivalent to the transition probabilities from possible current health-states of the structure to future health-states.

The current paper aims to demonstrate how such models may be determined using information gathered during the operational evaluation stage of the structural health monitoring paradigm. This information may include knowledge of the operational and environmental conditions under which the structure will operate, in addition to initial physics-based modelling of the structure. A probabilistic transition model describing the degradation of a four-bay truss is developed here, with finite element simulation used to yield knowledge of the load paths within the structure when it is in differing health-states. The paper concludes with a discussion of the importance of probabilistic degradation models within SHM decision-making. The discussion highlights the challenges that arise due to the lack of data available prior to the implementation of an SHM system and suggests for how these may be overcome.

Keywords Risk · Decision-making · Probabilistic graphical models · Degradation modelling

1 Introduction

Structural health monitoring (SHM) is a field of engineering that is concerned with damage detection in structures and infrastructure via the development and implementation of data acquisition and processing systems [1]. A key motivation for the development and implementation of SHM systems is to facilitate the decision-making processes associated with the operation and management of high-value or safety-critical assets. One approach to decision-making in the context of SHM is by the use of a probabilistic risk-based framework based upon probabilistic graphical models (PGMs) [2], in which actions on, and failure modes of the structure are assigned costs and optimal decisions are made through the maximisation of expected utility gain, or the minimisation of expected utility loss.

An agent tasked with making decisions regarding the operation and management of a structure may utilise health-state information inferred via an SHM system to make better informed and more optimal decisions. However, given solely information regarding the structural health-state at the current instance in time, the agent may only make well-informed decisions ad hoc. In order to make well-informed decisions on policies that include preventative actions, the agent requires information about the future health-states of the structure. This information can be gained by developing transition models that forecast future health-states given the current health-state and each decidable action. For the case that the decided action is 'do nothing' the health-state transition model will forecast the degradation of a structure.

Degradation models of differing complexities have been used within the field of engineering for reliability assessment, maintenance planning and prognosis [3]. In general, the models can be categorised in terms of a combination of the following

A. J. Hughes (✉) · R. J. Barthorpe · K. Worden
Dynamics Research Group, Department of Mechanical Engineering, University of Sheffield, Sheffield, UK
e-mail: ajhughes2@sheffield.ac.uk; r.j.barthorpe@sheffield.ac.uk; k.worden@sheffield.ac.uk

K. Grimmelsman (ed.), *Dynamics of Civil Structures, Volume 2*, Conference Proceedings of the Society for Experimental Mechanics Series, https://doi.org/10.1007/978-3-030-77143-0_5

criteria; physics-based or data-based, deterministic or probabilistic and continuous state or discrete state. A commonly used degradation model is Paris' law for crack growth given by the following equation,

$$\frac{da}{dN} = C(\Delta K)^m \tag{1}$$

where a is the crack length, N is the load cycle, ΔK is the stress intensity range and C and m are constants. After a little thought one can reason that Eq. (1) is a deterministic, physics-based model of a continuous state. Different categories of degradation model are applicable in different scenarios depending on the context. For example, in a situation where little is known of the underlying physics governing the degradation, but data are readily available, one may opt for a data-based model. Conversely, if the physics are known but data availability is low, a physics-based model may be more suitable. Whether continuous or discrete states are modelled also depends on the nature of the application; considerations for this include the required model fidelity and the computational cost/time. Without delving too far into metaphysics, it is reasonable to assert that, in general, the future is inherently uncertain. For this reason, with regard to the use of deterministic versus probabilistic models, the latter have a distinct advantage as they are capable of representing uncertainty. Fortunately, many deterministic degradation models can be used to obtain probabilistic outputs via methods such as sequential Monte Carlo sampling [4].

In the context of SHM and decision-making, a variety of health-state transition models have been employed. In [5], a probabilistic interpretation of Paris' law is used to develop a degradation model in a maintenance decision process for a simulated wind turbine tower. In [6], a continuous health-state variable is given nonlinear Gaussian transition models in a partially observable Markov decision process (POMDP) based on a normalised unscented Kalman filter; this approach has the property that there is a non-zero probability that the health-state transitions to a less-damaged state, meaning that the structural degradation is not strictly monotonic. In [7], qualitative data obtained from the inspection of mitre gate components is used to derive a health-state transition matrix for a Markovian decision process for optimal maintenance decisions.

The current paper aims to present a general methodology for determining a health-state transition matrix for use in a probabilistic risk-based decision paradigm for the operation and maintenance of structures as developed in [2]. The methodology will be demonstrated using a case study of a four-bay truss. Finally, the importance of health-state transition models within the risk-based decision framework will be discussed, and the challenges associated with their development will be highlighted.

2 Probabilistic Risk-Based SHM

The approach proposed in [2] facilitates decision-making in the context of SHM by incorporating aspects of probabilistic risk assessment into a probabilistic graphical model framework. For brevity, here, a short introduction to probabilistic graphical models is provided, followed by a summary of the risk-based decision framework; for a more comprehensive explanation, the reader is directed to the original paper.

2.1 Probabilistic Graphical Models

Probabilistic graphical models are graphical representations of factorisations of joint probability distributions and are a powerful tool for reasoning and decision-making under uncertainty. For this reason, they are apt for representing and solving decision problems in the context of SHM, where there is uncertainty in the health-states of structures. While there exist multiple forms of probabilistic graphical model, the key types utilised for the risk-based decision frameworks are Bayesian networks (BNs) and influence diagrams (IDs) [8].

Bayesian networks are directed acyclic graphs (DAGs) comprised of nodes and edges. Nodes represent random variables and edges connecting nodes represent conditional dependencies between variables. In the case where the random variables in a BN are discrete, the model is defined by a set of conditional probability tables (CPTs). For continuous random variables, the model is defined by a set of conditional probability density functions (CPDFs).

Figure 1 shows a simple Bayesian network comprised of three random variables X, Y and Z. Y is conditionally dependent on X and is said to be a child of X while X is said to be a parent of Y. Z is conditionally dependent on Y and can be said to be a child of Y and a descendant of X while X is said to be an ancestor of Z. The factorisation described by the Bayesian network shown in Fig. 1 is given by $P(X, Y, Z) = P(X) \cdot P(Y|X) \cdot P(Z|Y)$. Given observations on a subset of nodes

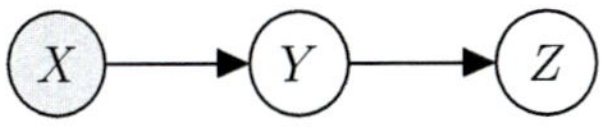

Fig. 1 An example Bayesian network

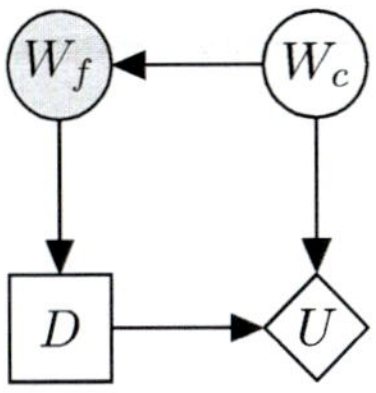

Fig. 2 An example influence diagram representing the decision of whether to go outside or stay in under uncertainty in the future weather condition given an observed forecast

in a BN, inference algorithms can be applied to compute posterior distributions over the remaining unobserved variables. Observations of random variables are denoted in a BN via grey shading of the corresponding node, as is demonstrated for X in Fig. 1.

Bayesian networks may be adapted into influence diagrams to model decision problems. This augmentation involves the introduction of two additional types of node that are shown in Fig. 2: decision nodes, denoted as squares, and utility nodes, denoted as rhombi. For influence diagrams, edges connecting random variables to utility nodes denote that the utility function is dependent on the states of the random variables. Similarly, edges connecting decisions nodes to utility nodes denote that the utility function is dependent on the decided actions. Edges from decision nodes to random variable nodes indicate that the random variables are conditionally dependent on the decided actions. Edges from random variable or decision nodes to other decision nodes do not imply a functional dependence but rather order, i.e. that the observations/decisions must be made prior to the next decision being made.

To gain further understanding of IDs, one can consider Fig. 2. Figure 2 shows the ID for a simple binary decision; stay home and watch TV or go out for a walk, i.e. $domain(D) = \{\text{TV, walk}\}$. Here, the agent tasked with making the decision has access to the weather forecast W_f which is conditionally dependent on the future weather condition W_c. The weather forecast and future condition share the same possible states $domain(W_f) = domain(W_c) = \{\text{bad, good}\}$. The utility achieved, U, is then dependent on both the future weather condition and the decided action. For example, one might expect high utility gain if the agent decides to go for a walk and the weather condition is good.

In general, a policy δ is a mapping from all possible observations to possible actions. The problem of inference in influence diagrams is to determine an optimal strategy $\boldsymbol{\Delta}^* = \{\delta_1^*, \ldots, \delta_n^*\}$ given a set of observations on random variables where δ_i^* is the ith decision to be made in a strategy $\boldsymbol{\Delta}^*$ that yields the *maximum expected utility* (MEU). Defined as a product of probability and utility, the expected utility can be considered as a quantity correspondent to risk.

2.2 Decision Framework

A probabilistic graphical model for a general SHM decision problem across a single time-slice is shown in Fig. 3. Here, a maintenance decision d is shown for a simple fictitious structure $\boldsymbol{S}$, comprised of two substructures s_1 and s_2, each of which is comprised of two components; c_{1-2} and c_{3-4}, respectively.

The overall decision process model shown in Fig. 3 is based upon a combination of three sub-models; a statistical classifier, a failure-mode model and a transition model.

Within the decision framework, a random variable denoted $\boldsymbol{H}_t$ is used to represent the latent global health-state of the structure at time t. For this decision process, a posterior probability distribution over the latent health-state $\boldsymbol{H}_t$ is inferred via observations on a set of discriminative features $\boldsymbol{v}_t$. It is assumed that the generative conditional distribution $P(\boldsymbol{v}|\boldsymbol{H})$ is learned implicitly or explicitly, depending on the choice of statistical classifier.

The failure condition of the structure F_S is represented as a random variable within the PGM and is conditionally dependent on the health-states of the substructures denoted by the nodes hs_1 and hs_2. The health-states of the substructures are dependent on the local health-states of the constituent components denoted by the nodes hc_{1-4}. The local health-states of the components are summarised in the global health-state vector $\boldsymbol{H} = \{hc_1, hc_2, hc_3, hc_4\}$. The conditional probability tables defining the relationship between random variables correspond to the Boolean truth tables for each of the logic gates in the fault tree defining the failure mode F_S [9, 10]. This failure-mode model is repeated in each time-step. The failure states associated with the variable F_S are given utilities via the function represented by the node U_F. As it is necessary to consider the future risk of failure in the decision process, these utility functions are also repeated for each time-step.

Figure 4 shows the influence diagram of the transition sub-model. By interpretation of the graphical model shown in Fig. 4, one can realise that the transition sub-model is solely formed of the conditional probability distribution $P(\boldsymbol{H}_{t+1}|\boldsymbol{H}_t, d_t)$

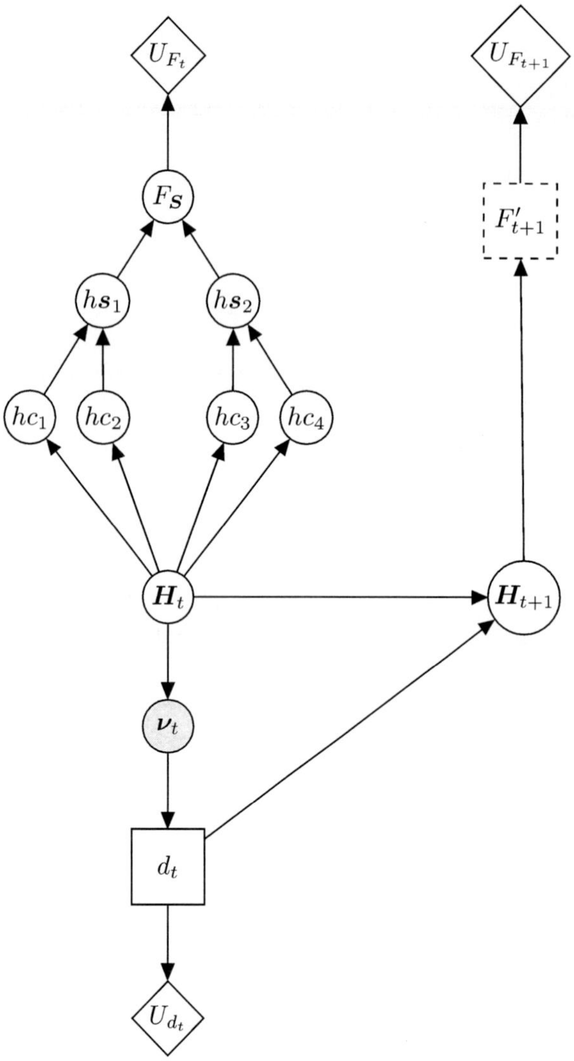

Fig. 3 An influence diagram representing the partially observable Markov decision process over one time-slice for determining the utility-optimal maintenance strategy for a simple structure comprised of four components. The fault tree failure-mode model for time $t+1$ has been represented as the node F'_{t+1} for compactness

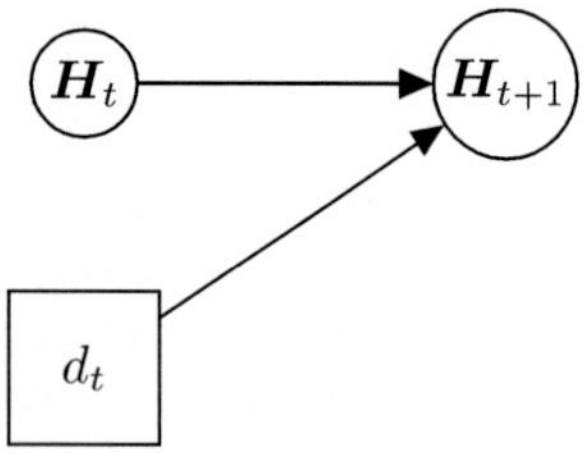

Fig. 4 An influence diagram representing the transition sub-model of the overall SHM decision process

and that the future health-state $\boldsymbol{H}_{t+1}$ is dependent only on the current health-state and the action decided in the current time-slice. An underlying assumption of the decision framework presented in [2], that facilitates the modelling process, is that structures can be represented as a hierarchical combination of discrete substructures/regions. A consequence of this assumption is that the health-states of interest are all represented as discrete random variables, hence, the transition models required are matrices. For a given decided action a, and assuming a finite number N of possible discrete global health-states, the conditional probability table $P(\boldsymbol{H}_{t+1}|\boldsymbol{H}_t, d_t = a)$ is given by an $N \times N$ square matrix whose i, jth entry is the probability of transitioning from the ith to the jth health-state and $i, j \in \mathbb{Z} : 1 \leq i, j \leq N$. Additionally, it is assumed that the Markov decision process is stationary, i.e. $P(\boldsymbol{H}_{t+1}|\boldsymbol{H}_t, d_t = a)$ is invariant with respect to t. Because of this stationarity, assuming no intervention is made (d_t = 'donothing' $\forall\, t$), the future global structural health-state is forecast as,

$$P(\boldsymbol{H}_{t+n}) = P(\boldsymbol{H}_t) \cdot P(\boldsymbol{H}_{t+1}|\boldsymbol{H}_t, d_t = \text{'do nothing'})^n \tag{2}$$

where n is the number of discrete time-slices forecast over, and $P(\boldsymbol{H}_t)$ and $P(\boldsymbol{H}_{t+n})$ are $1 \times N$ multinomial probability distributions over the global health-states at times t and $t+n$, respectively.

3 Developing Transition Models for Risk-Based SHM

As with the established paradigm for conducting an SHM campaign (detailed in [1]), the risk-based approach is formed of several distinct stages. The risk-based approach consists of: operational evaluation, failure-mode modelling, decision modelling, data acquisition, feature selection and statistical modelling. Most crucial to the development of transition models is the operational evaluation stage. The current section outlines the information that must be obtained for the development of transition models, provides discussion around the quantification of the uncertainty in operational conditions and offers an explanation of how the quantified uncertainty may be used in conjunction with a physics-based model to develop transition models.

3.1 Operational Evaluation

The operational evaluation stage, for both the traditional and probabilistic risk-based structural health monitoring paradigms, seeks to assess the context in which a structural health monitoring campaign is to be conducted. It is during this stage that the operational and environmental conditions for the structure of interest are considered. Furthermore, failure modes of interest are determined and key health-states of the structure identified.

For the development of transition models in the probabilistic risk-based approach, during the operation evaluation stage, it is necessary to identify factors that will influence the way in which the structure will degrade. Many of these factors may be specific to the type of structure on which SHM is being conducted. Information regarding the operational conditions that must be obtained includes the anticipated forcing amplitudes, locations and temporal variations. These operational conditions will influence the fatigue life of the structure. Environmental conditions are also important to consider. Examples of important environmental factors include operating temperatures and the presence/absence of water. The anticipated operational temperature ranges are important to consider as these potentially introduce thermally induced stresses in addition to other temperature effects on material properties such as fracture toughness. Furthermore, whether the structure will be in the presence of water is a key factor as this may introduce structural degradation mechanisms such as corrosion and erosion. An important consideration to make when considering operational and environmental conditions is that degradation mechanisms may interact with one another. A notable example of this effect occurring is within the core of light-water nuclear reactors where stainless steel structural components experience accelerated brittle fracturing as a result of interplay between multi-physical phenomena in a process known as irradiation-assisted stress corrosion cracking (IASCC) [11].

With the operational and environmental conditions of the structure considered and potential degradation mechanisms determined, the failure modes of interest for the structure and critical substructures, components and joints can be identified. Subsequently, it is important to define damage for each critical substructure, component and joint, i.e. the possible local health-states. Depending on factors such as materials and local operational and environmental conditions, different components/joints may be susceptible to different types of damage; for example, composite components may experience delamination, whereas metallic components may experience fatigue cracking. For each component, criteria for each of the relevant failure mechanisms should be specified.

Irrespective of the type of damage associated with each component/joint, it is reasonable to assert that the discrete random variables corresponding to the local health-states will have a cardinality of at least 2. In the most simple case, each local health-state variable could possess states corresponding to 'undamaged' and 'failed', where the 'failed' state represents the component being unfit-for-purpose. In some scenarios, it may also be desirable to consider extents of damage and the functionality of the component/joint at varying damage extents. Some components/joints may possess health-states associated with the presence of damage whilst continuing to function at their full, or partial capacity. Although these states are not necessarily associated with any immediate risk with regard to the failure of the global structure, they may still be important to consider as they may increase the propensity for transitioning to other more advanced damage states that do have high risk associated. An example of a component that may require this consideration is a load-bearing structural member in which partial thickness cracks may form.

3.2 Handling Uncertainty

For most applications of structural health monitoring, perfect knowledge of the operational and environmental conditions will not be available prior to the implementation of the system. It is for this reason, that uncertainties should be considered and quantified where possible. While there exists a number of methodologies for the quantification of uncertainty, including interval analysis and Dempster-Shafer theory [12, 13], here, it is considered reasonable to continue using probability theory for consistency with the probabilistic risk-based decision framework.

For each of the key environmental and operation conditions, statistical distributions quantifying the ranges, likely values and/or variance in the conditions should be elicited from an expert judgement, and where possible, observed data. In a Bayesian setting, these distributions may be updated as measurements are collected, and the transition models re-estimated.

3.3 Generating Transition Models

To generate the degradation transition models, a physics-based model is required. The function of the model is to simulate the structure and specifically its critical components in each of the global health-states and under specified operational and environmental conditions. The simulated structure can then be evaluated with respect to the failure criteria identified in the operational evaluation stage to determine whether state transitions occur.

With respect to modelling the degradation of a structure, the purpose of the physics-based model is to determine a distribution over the quantities of interest in which the failure criteria are specified, conditioned on the uncertain operational and environmental conditions. In the case that the physics-based model employed is inherently stochastic (such as a probabilistic fracture mechanics model), this conditional distribution may be determined analytically. In the case that the physics-based model employed is deterministic (such as a finite element model), this distribution may be determined by applying sampling methods to the probability distributions for the operational and environmental conditions, and querying the physics-based model accordingly.

Once a distribution over the quantities of interest has been determined, a distribution over local failure events can be produced by executing the logical operations defining the failure criteria. Again, this distribution is conditioned on the operational and environmental conditions. This conditional distribution over local failure events can then be mapped into transitions in the global health-state by utilising the definition of $\boldsymbol{H}$ as a vector containing the local health-states of the critical components, joints and substructures.

At this stage, it is necessary to marginalise out the variable operational and environmental conditions to obtain the distribution $P(\boldsymbol{H}_{t+1}|\boldsymbol{H}_t, d = 0)$. Additionally, to ensure a valid probability distribution is produced, normalisation should be carried out.

Developing transition models for specific actions (such as repairs) is typically a problem that is highly dependent on the context.

4 Case Study: Four-Bay Truss

To demonstrate how probability distributions quantifying uncertainty in operational conditions may be used in conjunction with a physics-based model to generate a transition model for a risk-based SHM decision process, the methodology was applied to a case study of a physical four-bay truss structure identical to that used in [14], and shown in Fig. 5. The truss was composed of 20 aluminium members, each with a cross-sectional area of 177 mm^2. The horizontal and vertical members of the truss possessed lengths of 250 mm, resulting in the overall structure having a length of 1 m and a height of 0.25 m. The members were pinned together using steel bolts in lubricated holes. For illustrative purposes, fictitious operational conditions were assumed.

To avoid obfuscating the development of the transition model, it was elected to ignore the failure of joints and the horizontal and vertical members and instead focus on the failures of the cross-members. Denoting the local health-states of the eight cross-members as hm_9 to hm_{16}, the global health-state of the structure can be expressed as the vector $\boldsymbol{H} = \{hm_9, \ldots, hm_{16}\}$. Additionally, for the purposes of demonstration, binary health-states for each of the 8 cross-members were considered resulting in 256 possible global health-states. From hereon in, a convenient referencing scheme for the global health-states is adopted where $\boldsymbol{H}$ is given a superscript corresponding to the decimal representation of the 8-bit binary number (with ascending powers of two from left to right) specified by the vector $\boldsymbol{H}$, i.e. the undamaged health state

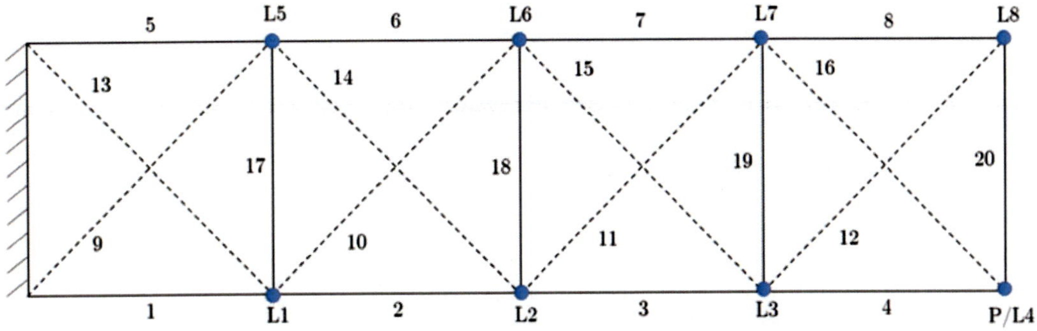

Fig. 5 A two-dimensional four-bay truss comprised of 20 members, eight of which are removable and denoted by a dashed line. Loads are applied at points L, and a preload is applied at point P. Load positions are shown as blue dots. The bays are numbered left to right from 1 to 4

$\boldsymbol{H} = \{0, 0, 0, 0, 0, 0, 0, 0\}$ is denoted as $\boldsymbol{H}^0$, and the health-state corresponding to the failure of the cross-members in the first bay $\boldsymbol{H} = \{1, 0, 0, 0, 1, 0, 0, 0\}$ is denoted as $\boldsymbol{H}^{17}$.

Finally, a binary decision d was considered for the structure, with possible courses of action 'do nothing' and 'perform maintenance'; for conciseness, these actions will be denoted with $d = 0$ and $d = 1$, respectively. In this case study, it is assumed that the 'perform maintenance' action is equivalent to the replacement of all cross-members with the structure consequently returned to its undamaged state.

4.1 Operational Conditions

Operational conditions were assumed for the structure such that the stress experienced in cross-members has a degree of stochasticity. Specifically, it was assumed that there would be uncertainty in both the load and the location that the load is applied to the structure at each time-step. In addition to the variable load, a constant preload of 5 kg was applied to the structure at point P.

The magnitude of the load w was assumed to vary in accordance with the discrete uniform distribution,

$$w \sim \mathcal{DU}(0, w_{max}; n) \tag{3}$$

where w_{max} was determined such that $P(\boldsymbol{H}^0_{t+1} | \boldsymbol{H}^0_t, d_t = 0) = 0.8$ and each load magnitude had probability of $P(w) = \frac{1}{n}$ with $n = 100$.

The position of the load was also assumed to vary according to a discrete uniform distribution over 8 candidate locations labelled L1 to L8 in Fig. 5. This distribution may be formalised as:

$$L \sim \mathcal{DU}(1, 8) \tag{4}$$

Hence, the operational conditions can be summarised as a vector $\boldsymbol{c}_o = \{w, L\}$. In total, 800 possible operational conditions were considered.

4.2 Failure Criteria

For each cross-member, three modes of failure were considered; yielding under tension, buckling under compression, and supercritical crack growth.

A cross-member was considered to have failed by yielding, if the tensile stress in the member exceeded the ultimate tensile stress of aluminium, where $\sigma_{UTS} = 300$ MPa. The event of a cross-member m_i failing via yielding is denoted as Y_i.

A cross-member was considered to have failed by buckling when the compressive stress within a member exceeded the buckling stress σ_b. The critical buckling stress for a slender beam is given by the following equation [15],

$$\sigma_b = \frac{\pi^2 EI}{A(KL)^2} \tag{5}$$

where E is the Young's modulus, I is the cross-sectional second moment of area, A is the cross-sectional area, K is the effective length factor and is dependent on the boundary conditions, and L is the length of the member. As the truss was constructed in a way that allows in-plane rotation at the ends of each member, a pinned-pinned boundary condition was assumed, resulting in an effective length factor of $K = 1$. Taking the Young's modulus of aluminium to be $E = 70$ GPa, the critical buckling stress was found to be a compressive stress of $\sigma_b = 270$ MPa. The event of a cross-member m_i failing via buckling is denoted as B_i.

The final failure method considered for the cross-members was supercritical crack growth. For this failure mechanism, it was assumed that each member possessed a crack in the centre across the entire width of the member and at the midpoint along the length with probability 0.1. The size of the crack in meters was assumed to be continuous uniformly distributed according to,

$$2a \sim \mathcal{U}(0, b) \tag{6}$$

where $2a$ is the crack size and $b = 0.0125$ and is the half width of the cross-members.

Assuming the cross-members can be modelled as a finite plate and with plane strain conditions, the mode I stress intensity factor K_I for a cracked member can be given by the following equation [16],

$$K_I = G\sigma\sqrt{\pi a} \tag{7}$$

where σ is the applied stress, and G is a geometric factor given by,

$$G = \frac{1 - \frac{a}{2b} + 0.326(\frac{a}{b})^2}{\sqrt{1 - \frac{a}{b}}} \tag{8}$$

A cracked cross-member was considered to have failed when the stress intensity factor exceeded the critical stress intensity factor K_c. For the aluminium members, it was taken that $K_c = 24\,\text{MPa} \cdot \text{m}^{\frac{1}{2}}$. The event of a cross-member m_i failing via supercritical cracking is denoted as C_i.

The initial variable structural conditions can be summarised in a vector $\boldsymbol{c}_s = \{2a_9, \ldots, 2a_{16}\}$, where $2a_i$ is the crack length present in cross-member m_i. Here, it should be noted that the $\boldsymbol{c}_s$ is considered independently of $\boldsymbol{H}$.

4.3 Transition Modelling

To determine the stresses within the structure under the variable operational and structural conditions, a finite element model of the truss was developed. The finite element model was validated with a set of strain measurements taken from the physical truss in its undamaged condition.

A wrapper function was produced to iterate over the global health-states $\boldsymbol{H}_t$. Additionally, the function was used to generate random samples $\boldsymbol{c}^*$ from the probability distributions specifying the uncertain operational and structural conditions $\boldsymbol{c} = \{\boldsymbol{c}_o, \boldsymbol{c}_s\}$. Afterwards, the function queried the finite element model to obtain the stresses in the cross-members for the given global health-state and a random sample of operational and structural conditions.

Asserting $d = 0$, for an initial global health-state $\boldsymbol{H}_t$ and a randomly sampled set of conditions $\boldsymbol{c}^*$, a health-state transition was defined as $\boldsymbol{H}_{t+1} = \boldsymbol{H}_t + \delta\boldsymbol{H}$ where $\delta\boldsymbol{H} = \{\delta hm_9, \ldots, \delta hm_{16}\}$ is an 8-bit binary vector and,

$$\delta hm_i = \mathbb{1}[(Y_i \vee B_i \vee C_i)|\boldsymbol{H}_t, d = 0, \boldsymbol{c}^*] \tag{9}$$

where $\mathbb{1}$ denotes the indicator function and $\vee$ denotes the inclusive-or logical operator. Here, Eq. (9) corresponds to evaluating cross-member failures with respect to the previously discussed criteria for yielding, buckling and cracking. Subsequently, the conditional probability of transitioning from $\boldsymbol{H}_t^i$ to $\boldsymbol{H}_{t+1}^j$ given $\boldsymbol{c}^*$ was specified such that,

$$P(\boldsymbol{H}_{t+1}^{j}|\boldsymbol{H}_{t}^{i}, d=0, \boldsymbol{c}^{*}) = \begin{cases} 1 & \text{if } \delta\boldsymbol{H} = \boldsymbol{H}_{t+1}^{j} - \boldsymbol{H}_{t}^{i} \\ 0 & \text{otherwise.} \end{cases} \tag{10}$$

To populate the transition matrix $P(\boldsymbol{H}_{t+1}|\boldsymbol{H}_t, d = 0)$, the variability in the conditions $\boldsymbol{c}$ must be marginalised out and the distribution normalised. This was achieved by calculating the i, jth entry of the transition matrix as,

$$P(\boldsymbol{H}_{t+1}^{j}|\boldsymbol{H}_{t}^{i}, d=0) = \frac{\sum_{1}^{N_s} P(\boldsymbol{H}_{t+1}^{j}|\boldsymbol{H}_{t}^{i}, d=0, \boldsymbol{c}^{*})}{N_s} \tag{11}$$

where N_s is the number of queries of the finite element model per $\boldsymbol{H}_t$.

The transition model for the action corresponding to 'do nothing' was estimated with the described procedure using $N_s = 10^4$. The heatmap of the resulting transition matrix $P(\boldsymbol{H}_{t+1}|\boldsymbol{H}_t, d = 0)$ is shown in Fig. 6. A dominant lighter colour line can be seen along the diagonal in Fig. 6; this indicates that the structure has a tendency to remain in the same health-state over a single time-step. Furthermore, it can be seen that the elements in the lower-right triangle of the graph (which corresponds to the lower-left triangle of the transition matrix) consists entirely of zero elements; a result of the implicit constraint imposed through Eqs. (9) and (10) that the structure monotonically degrades. Taking the $\log_{10}$ of the conditional probability distribution (with an offset of $+0.01$ so that zero elements may be plotted with finite values) reveals further structure in the transition matrix as lower probability transitions are made more visible, as can be seen in Fig. 7. Figure 7 shows that the transition matrix has fractal pattern akin to the Sierpiński triangle. Due to the fact that the global health-state is represented as an 8-bit binary vector, the set of all allowable transitions assuming only monotonic degradation (i.e. once a bit is 'turned on', it cannot be 'turned off'), form a Sierpiński triangle [17]. The possible transitions shown in Fig. 7 are a subset of the Sierpiński triangle with some elements missing due to physical effects disallowing some transitions; for example, if the truss were to collapse due to the failure of the first bay, then the members in the other bays would no longer be able to fail as the structure would cease to support the load.

For completeness, the transition matrix for the 'perform maintenance' action $P(\boldsymbol{H}_{t+1}|\boldsymbol{H}_t, d = 1)$, was specified by making the assumption that the replacement of all cross-members returns the structure to the undamaged health-states, as shown by the following function:

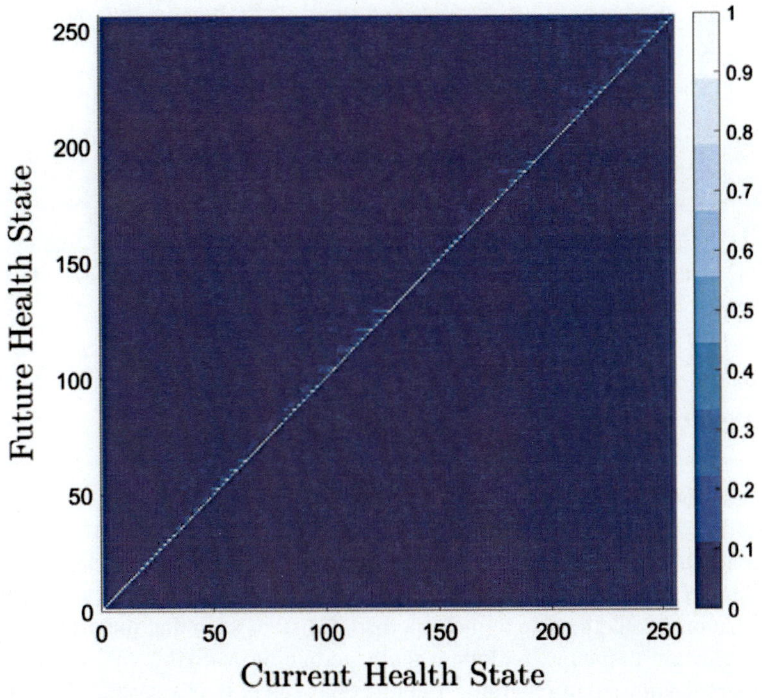

Fig. 6 A heatmap showing the transition matrix $P(\boldsymbol{H}_{t+1}|\boldsymbol{H}_t, d = 0)$

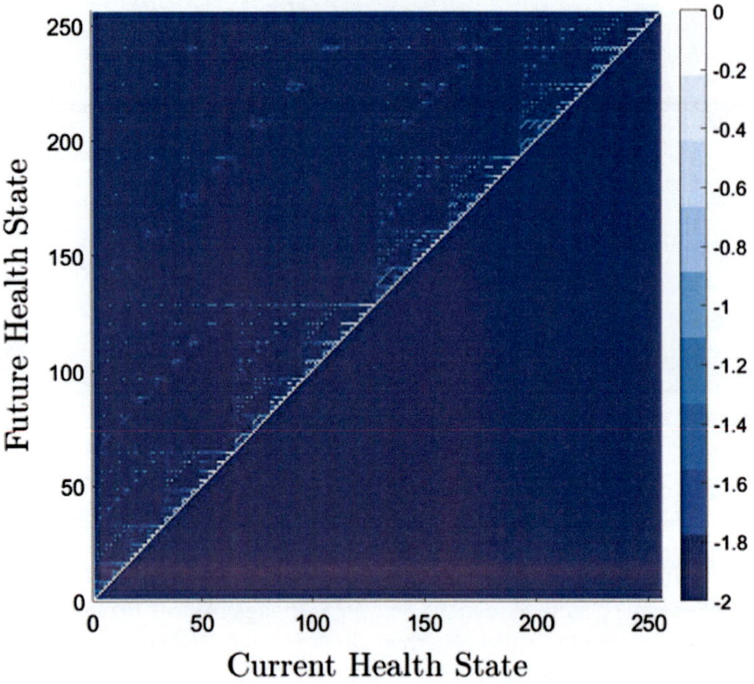

Fig. 7 A heatmap showing the log probability of the transition matrix with an offset, $\log_{10}(P(\boldsymbol{H}_{t+1}|\boldsymbol{H}_t, d = 0) + 0.01)$

$$P(\boldsymbol{H}^j_{t+1}|\boldsymbol{H}^i_t, d = 1) = \begin{cases} 1 & \text{for } j = 0 \\ 0 & \text{otherwise.} \end{cases} \tag{12}$$

The current section has demonstrated a sensible methodology for developing a health-state transition model for a structure by means of a case study. The next steps would be to evaluate and test the transition model, though this is omitted here as it is outside the scope of the current paper.

5 Discussion

The current sections aim to highlight and discuss the importance of health-state transition models in the context of risk-based decision-making for SHM and for the specific problem of prognosis. Additionally, discussion will be made around the challenges associated with the development of the transition models.

5.1 Importance of Transition Models

In general, when it comes to decision-making, possessing information or beliefs regarding future events/states is crucial. This statement becomes most apparent when taking this notion *ad absurdum*. At one extreme, if one possesses no information or belief regarding future events/states, then there is no reason for one to expect that any single course of action is better than any other. At the other end, if one somehow becomes clairvoyant and possesses perfect information regarding future events/states, then it follows that one would be able to make perfect decisions such that maximum rewards may be reaped.

As it happens, almost all decision problems, including those pertaining to SHM, fall somewhere between these two extremes, where belief and partial knowledge regarding future events/states is possessed. Nonetheless, in the context of SHM, increased expected utility gain provides a strong argument for striving towards improved knowledge regarding future health-states by the development of transition models.

In addition to allowing closer to optimal decisions to be made within the risk-based framework, a good transition model allows for a pseudo-prognosis for the structure to be made by utilising Eq. (2). By propagating the belief in the current health-state forward in time according to Eq. (2), and by evaluating the risk of failure associated with the predicted distribution over future health-states, at each time-step until the risk exceeds the cost of one of the candidate courses of action, one can obtain an estimate for the anticipated number of time-steps until an action should be taken. Whilst this result is not as powerful as a true-prognosis that yields remaining useful life, this information is still beneficial as it provides the expected time available to execute a course of action.

5.2 Challenges

There are numerous challenges associated with the development of transition models.

A primary challenge pertains to the validation of transition models. For many applications of SHM, the monitoring campaign will be for a newly built structure from which data are yet to be acquired at the time that the transition model must be developed. Without any observed state transitions to validate the model, one must rely solely on prior knowledge of the underlying physics that govern the degradation. One possible option is to independently validate the physics-based models used to develop the transition model via hybrid testing, or performing experiments on individual components or substructures. Alternatively, in situations where an SHM system is being retrofitted to an existing structure there may be historical data detailing health-state transitions that may be used to validate the degradation model.

The issue of validation is further complicated if the structure of interest is unique. For such a structure, even in a scenario where one is able to update the transition model with observed state transitions, it is possible, and in many cases likely, that only a small subset of the total possible state transitions will be observed throughout the operational lifetime; thereby leaving potentially large portions of the transition model without validation. In the context of population-based SHM [18–20], a single transition model may be applied to all members of a fleet of homogeneous structures and also updated with state transitions observed from each instance of the structure. The process of continually validating transition models online may be achieved through active learning [21].

Another challenge is the cost, both in terms of money and time, associated with the development of transition models. The development cost of a transition model will depend highly on the complexity of the structure for which a model is being developed, and the range of operational and environmental conditions that must be considered. For complex structures, the high-fidelity models capable of the multi-physics that may be required to simulate all the necessary failure mechanisms to develop a transition model are expensive and time-consuming to develop, often requiring teams of highly skilled engineers. The financial argument for the development of such models should be constructed and evaluated during the operational evaluation stage of the SHM process, taking into account whether the structure is of high-value, or safety-critical.

The computational cost of the development and implementation of the transitional model should also be considered. During the development of the transition model, it is possible that a physics-based model is queried numerous times. For complex structures, and high-fidelity models these simulations required large computing times. As the number of influential operating and environmental conditions increases, the number of samples required to adequately cover the input space will also increase. Taking this factor into account with the possibility that high-fidelity physics-based simulations may need to be queried many times, the calculation of the transition models may have prohibitively long computation times. A possible solution to this issue would be to use a surrogate model, where an interpolation function that is relatively cheap to query is trained on a subset of the outputs of the physics-based model.

Finally, a challenge pertaining to maintenance action transition models is left as an open topic for research and discussion. In a few limited cases, such as when repair corresponds to replacement of all failed components (as is assumed for the case study in the current paper), it may be reasonable to assume that the structure returns to its original undamaged case. However, in general, for less extreme and more realistic approaches to structural repair, this does not hold and, in fact, it is possible that the state to which the structure transitions was not considered during the original development of the transition model [22]. Here, the challenge lies with determining reasonable assumptions that allow one to avoid redeveloping the transition model after every intervention, or to conceive of methods for adapting the health-states considered within the risk-based decision framework.

6 Summary

The aim of the current paper has been to present a general methodology for developing structural health-state transition models for use in a probabilistic risk-based decision framework for SHM. Using a four-bay truss for a case study, a degradation model in the form of a probabilistic transition matrix was developed by considering uncertain operational conditions in conjunction with a physics-based model. Finally, discussions were made focussing on the challenges with developing health-state transition models but also on the importance of the models for both the risk-based decision framework, and their application to the problem of prognosis in SHM.

Acknowledgments The authors would like to acknowledge the support of the UK EPSRC via the Programme Grant EP/R006768/1. KW would also like to acknowledge support via the EPSRC Established Career Fellowship EP/R003625/1.

References

1. Farrar, C.R., Worden, K.: Structural Health Monitoring: A Machine Learning Perspective. Wiley, Chichester (2013)
2. Hughes, A.J., Barthorpe, R.J., Dervilis, N., Farrar, C.R., Worden, K.: A probabilistic risk-based decision framework for structural health monitoring. Mech. Syst. Signal Process. **150**, 107339 (2021)
3. Shahraki, A.F., Yadav, O.P., Liao, H.: A review on degradation modelling and its engineering applications. Int. J. Perform. Eng. **13**, 299–314 (2017)
4. Corbetta, M., Sbarufatti, C., Manes, A., Giglio, M.: Real-time prognosis of crack growth evolution using sequential Monte Carlo methods and statistical model parameters. IEEE Trans. Reliab. **64**(2), 736–753 (2015)
5. Hovgaard, M.K., Brincker, R.: Limited memory influence diagrams for structural damage detection decision-making. J. Civil Struct. Health Monit. **6**(2), 205–215 (2016)
6. Schöbi, R., Chatzi, E.N.: Maintenance planning using continuous-state partially observable Markov decision processes and non-linear action models processes and non-linear action models. Struct. Infrastruct. Eng. **12**(8), 977–994 (2016)
7. Vega, M.A., Todd, M.D.: A variational Bayesian neural network for structural health monitoring and cost-informed decision-making in miter gates. Struct. Health Monit. (2020). https://doi.org/10.1177/1475921720904543
8. Sucar, L.E.: Probabilistic Graphical Models: Principles and Applications. Springer, London (2015)
9. Bobbio, A., Portinale, L., Minichino, M., Ciancamerla, E.: Improving the analysis of dependable systems by mapping Fault Trees into Bayesian Networks. Reliab. Eng. Syst. Saf. **71**(3), 249–260 (2001)
10. Mahadevan, S., Zhang, R., Smith, N.: Bayesian networks for system reliability reassessment. Struct. Saf. **23**(3), 231–251 (2001)
11. Bruemmer, S.M., Simonen, E.P., Scott, P.M., Andresen, P.L., Was, G.S., Nelson, J.L.: Radiation-induced material changes and susceptibility to intergranular failure of light-water-reactor core internals. J. Nucl. Mater. **274**(3), 299–314 (1999)
12. Dempster, A.P.: Upper and lower probabilities induced by a multivalued mapping. Ann. Math. Stat. **38**(2), 325–339 (1967)
13. Shafer, G.: A Mathematical Theory of Evidence. Princeton University Press, Princeton, NJ (1976)
14. Worden, K., Ball, A.D., Tomlinson, G.R.: Fault location in a framework structure using neural networks. Smart Mater. Struct. **2**(3), 189–200 (1993)
15. Yoo, C.H., Lee, S.C.: Stability of Structures: Principles and Applications. Elsevier, Oxford (2011)
16. Anderson, T.L.: Fracture Mechanics: Fundamentals and Applications. Taylor and Francis, Boca Raton (2005)
17. Sloane, N.J.A.: The On-Line Encyclopedia of Integer Sequences: A001317. https://oeis.org/A001317. Accessed 08-12-2020
18. Bull, L.A., Gardner, P., Gosliga, J., Rogers, T.J., Dervilis, N., Cross, E.J., Papatheou, E., Maguire, A.E., Campos, C., Worden, K.: Foundations of population-based SHM, Part I: homogeneous populations and forms. Mech. Syst. Signal Process. **148**, 107141 (2021)
19. Gosliga, J., Gardner, P.A., Bull, L.A., Dervilis, N., Worden, K.: Foundations of Population-based SHM, Part II: Heterogeneous populations – graphs, networks, and communities. Mech. Syst. Signal Process. **148**, 107144 (2021)
20. Gardner, P., Bull, L.A., Gosliga, J., Dervilis, N., Worden, K.: Foundations of population-based SHM, Part III: Heterogeneous populations – mapping and transfer. Mech. Syst. Signal Process. 148, 107142 (2021)
21. Bull, L.A., Rogers, T.J., Wickramarachchi, C., Cross, E.J., Worden, K., Dervilis, N.: Probabilistic active learning: an online framework for structural health monitoring. Mech. Syst. Signal Process. **134**, 106294 (2019)
22. Gardner, P., Bull, L.A., Dervilis, N., Worden, K.: Overcoming the problem of repair in structural health monitoring: outlier-informed transfer learning. J. Sound Vib., 116245 (2021)

Cointegration for Structural Damage Detection Under Environmental Variabilities: An Experimental Study

J. C. Burgos, B. A. Qadri, and M. D. Ulriksen

Abstract An intricacy in vibration-based structural damage detection (VSDD) relates to environmental variabilities (EVs) imposing limitations to the damage detectability. One method that has been put forth to resolve the issue is cointegration. Here, non-stationary vibration features are linearly combined into stationary residuals, which are then employed as damage indices under the assumption that the non-stationarity is governed by environmental variabilities. In the present paper, the feasibility of using cointegration to mitigate environmental variabilities while retaining sensitivity to damage is examined through an experimental study with a steel beam. A temperature-based environmental variability is introduced to the beam by use of a heating cable, while damage is emulated by adding local mass perturbations. The vibration response of the beam in different environmental and structural states is captured and utilized as features in a cointegration-based damage detection scheme. The performance of the scheme is assessed and compared to that of a scheme not accounting for the variability on the basis of the false positive ratio (FPR), the true positive ratio (TPR), and the area under the receiver operating characteristic curve (AUC). The results show that cointegration effectively mitigates the temperature variability and allows for an improved damage detectability compared to that of the scheme without a mitigation strategy.

Keywords Structural health monitoring · Damage detection · Cointegration · Environmental variability · Experimental study

1 Introduction

The objective of vibration-based structural damage detection (VSDD) is to allow for a global assessment of whether or not the structural integrity of the system in question has been compromised. VSDD is conventionally resolved by comparing vibration features extracted from the current, potentially damaged structure to a baseline model composed of the corresponding vibration features extracted from the reference state of the structure [1]. An item that hinders the applicability of VSDD is variability in the environmental parameters (EPs), such as temperature, wind speed, and precipitation [2]. In particular, the dynamic properties of any structure are sensitive to changes in the EPs, thus vibration features exhibit significant changes over time also when no damage is present [3–5].

Numerous supervised and unsupervised learning methods have been proposed to mitigate the influence of environmental variabilities (EVs) in VSDD [1, 6]. The advantage of the unsupervised methods is that they do not require the EPs to be measured, hence making them easier to implement in practice. Cointegration, which is adapted from the field of econometrics [7], is one of these unsupervised methods [8]. The operating assumptions in cointegration-based damage detection are that the EVs render the vibration features non-stationary and that the effects of the EVs and damage do not couple [8]. Hereby, the non-stationary features can be linearly combined to one or multiple stationary residuals in which the non-stationary trends are purged [9].

In the present paper, we evaluate the application feasibility of the VSDD cointegration-based scheme proposed in [6, 10]. In particular, we apply the scheme to mitigate imposed EVs and hence allow for damage detection in a laboratory setting with a cantilevered beam. In the experiment, three damaged states are established by adding three different masses at the top of the beam, while temperature variations are imposed through a heating cable attached to the beam. The performance of the cointegration-based scheme is compared to that of a scheme that does not account for the EVs [11]. The false positive

J. C. Burgos (✉) · B. A. Qadri · M. D. Ulriksen
Department of Energy Technology, Aalborg University, Esbjerg, Denmark

K. Grimmelsman (ed.), *Dynamics of Civil Structures, Volume 2*, Conference Proceedings of the Society for Experimental Mechanics Series, https://doi.org/10.1007/978-3-030-77143-0_6

rate (FPR), the true positive rate (TPR), and the area under the receiver operating characteristic curve (AUC) are used as performance measures for the evaluation.

The paper is organized as follows: a methodology section outlines the explored damage detection schemes, followed by a description of the experimental setup. The damage detection results obtained using the two schemes are reported in the results section, and, lastly, the paper closes with some concluding remarks.

2 Methodology

Let $\boldsymbol{Y}_t \in \mathbb{R}^{F\times m}$ denote a vibration response matrix of a structural system measured through F sensors at a discrete observation $t \in [1,\ N]$. The rows of $\boldsymbol{Y}_t$ contain the F time series of the vibration response, $\boldsymbol{y}_{f,t} \in \mathbb{R}^{1\times m}$, where m is the number of time samples. From the response signals, a set of damage-sensitive vibration features are computed and subsequently used to characterize the state of the structural system as undamaged or damaged by use of some discordance measure.

2.1 *Mahalanobis Distance-Based Outlier Analysis of Response Covariances*

The covariance matrix of the vibration response matrix $\boldsymbol{Y}_t$ can be established from

$$\forall i \leq j \in [1,\ F]: \quad \boldsymbol{\Sigma}_{ij} = \frac{1}{m-1}\sum_{k=1}^{m}(\boldsymbol{Y}_{ik} - \mu_i)(\boldsymbol{Y}_{jk} - \mu_j), \tag{1}$$

where μ_i is the mean of the ith vibration response such

$$\forall i \in [1,\ F]: \quad \mu_i = \frac{1}{m}\sum_{k=1}^{m}\boldsymbol{Y}_{ik}. \tag{2}$$

In this scheme, the feature $\boldsymbol{x}_t$ is constructed from the n unique elements of the covariance matrix $\boldsymbol{\Sigma} \in \mathbb{R}^{F\times F}$, with $n = F(F+1)/2$ since $\boldsymbol{\Sigma} = \boldsymbol{\Sigma}^T$. Thus,

$$\boldsymbol{x}_t = [\boldsymbol{\Sigma}_{11}\ \boldsymbol{\Sigma}_{12}\ \ldots\ \boldsymbol{\Sigma}_{1F}\ \boldsymbol{\Sigma}_{22}\ \boldsymbol{\Sigma}_{23}\ \ldots\ \boldsymbol{\Sigma}_{2F}\ \ldots\ \boldsymbol{\Sigma}_{FF}]^T. \tag{3}$$

The Mahalanobis distance (MD) is used as discordance measure for $\boldsymbol{x}_t$. Let $\boldsymbol{\mu} \in \mathbb{R}^n$ and $\boldsymbol{S} \in \mathbb{R}^{n\times n}$ denote the mean vector and covariance matrix of the features from the training phase in the undamaged/reference structural state, then, in the testing phase, the deviation of $\boldsymbol{x}_t$ from the baseline model is quantified by

$$d_{M_t}^2 = (\boldsymbol{x}_t - \boldsymbol{\mu})^T \boldsymbol{S}^{-1} (\boldsymbol{x}_t - \boldsymbol{\mu}). \tag{4}$$

As such, $d_{M_t}^2$ constitutes the damage index from which inferences regarding the structural state are made. Specifically, if $\mathcal{D}$ is the selected threshold, then $d_{M_t}^2 \leq \mathcal{D}$ implies that the structure is undamaged, while $d_{M_t}^2 > \mathcal{D}$ implies that the structure is damaged.

2.2 *Cointegration-Based Scheme*

As mentioned introductory, we employ the cointegration-based scheme proposed in [6, 10]. Let $\boldsymbol{c}_{f,t} \in \mathbb{R}^m$ denote the real and imaginary parts of the unique Fourier coefficients of $\boldsymbol{y}_{f,t}$ and take

$$d_{M_{f,t}}^2 = \left(\boldsymbol{c}_{f,t} - \boldsymbol{\mu}_f\right)^T \boldsymbol{S}_f^{-1} \left(\boldsymbol{c}_{f,t} - \boldsymbol{\mu}_f\right), \tag{5}$$

where $\boldsymbol{\mu}_f \in \mathbb{R}^m$ and $\boldsymbol{S}_f \in \mathbb{R}^{m\times m}$ are the mean vector and covariance matrix of $\boldsymbol{c}_{f,t} \in \mathbb{R}^m$ computed in the training phase for N observations. We gather $d^2_{M_{f,t}}$ in $\boldsymbol{d}^2_{\boldsymbol{M}_f} \in \mathbb{R}^N$ and define the matrix $\boldsymbol{D}^2_{\boldsymbol{M}} = [\boldsymbol{d}^2_{\boldsymbol{M}_1}\ \boldsymbol{d}^2_{\boldsymbol{M}_2}\ \dots\ \boldsymbol{d}^2_{\boldsymbol{M}_F}]^T \in \mathbb{R}^{F\times N}$. The vector $\boldsymbol{d}^2_{\boldsymbol{M}_t}$, which corresponds to the columns of $\boldsymbol{D}^2_{\boldsymbol{M}}$, is used as feature in the cointegration.

The next step in the scheme is to find a stationary linear combination of the training features to construct a baseline model. To this end, the Johansen cointegration procedure (JCI) [12] is applied. In this procedure, the training features are fitted to the vector error-correction model (VECM)

$$\Delta\boldsymbol{d}^2_{\boldsymbol{M}_t} = \boldsymbol{\Pi}\boldsymbol{d}^2_{\boldsymbol{M}_{t\text{-}1}} + \sum_{i=1}^{k-1} \boldsymbol{C}_i \Delta\boldsymbol{d}^2_{\boldsymbol{M}_{t\text{-}i}} + \boldsymbol{\varepsilon}_t, \tag{6}$$

where $\Delta\boldsymbol{d}^2_{\boldsymbol{M}_t} = \boldsymbol{d}^2_{\boldsymbol{M}_t} - \boldsymbol{d}^2_{\boldsymbol{M}_{t\text{-}1}}$, $\boldsymbol{\varepsilon}_t \sim \mathcal{N}(0,\ \boldsymbol{S})$ is a noise vector, $\boldsymbol{\Pi}$ and $\boldsymbol{C}_i$ are $F \times F$ parameter matrices, and k is the model order (or number of included lags). If the VECM is a true error-correction model and the variables under consideration have the same order of integration (that is, $\Delta\boldsymbol{d}^2_{\boldsymbol{M}_t}$ and $\Delta\boldsymbol{d}^2_{\boldsymbol{M}_{t\text{-}i}}$ are stationary), then $\boldsymbol{\Pi}$ is rank deficient, say of rank r, and can therefore be decomposed as

$$\boldsymbol{\Pi} = \boldsymbol{A}\boldsymbol{B}^T, \tag{7}$$

where $\boldsymbol{A}, \boldsymbol{B} \in \mathbb{R}^{F\times r}$ are matrices spanning the column and row spaces of $\boldsymbol{\Pi}$. The column vectors in $\boldsymbol{B}$ can thus be used to project the non-stationary features into stationary residuals. According to the JCI, the vector that gives the most stationary residual is the particular column vector in $\boldsymbol{B}$ associated with the largest eigenvalue [13]. This vector, which we denote $\boldsymbol{\beta}$, is referred to as the cointegrated vector and its corresponding stationary residual, computed as

$$z = \boldsymbol{\beta}^T \boldsymbol{D}^2_{\boldsymbol{M}}, \tag{8}$$

is the cointegrated residual. A set of lags $k_j \in [k_1\ k_2\ \dots\ k_w]$ is tested in order to find the setting that gives the $\boldsymbol{\beta}$ yielding the most stationary z. The quantification of stationarity is conducted based on the Augmented Dickey-Fuller (ADF) test. The ADF test is employed to test the "degree of stationarity" of each z. In this test, the cointegrated residual is fitted to the model

$$\Delta z_t = \rho z_{t\text{-}1} + \sum_{j=1}^{q-1} b_j \Delta z_{t\text{-}j} + \varepsilon_t, \quad t \in [1,\ N], \tag{9}$$

where q is the number of lags (also referred to as model order) added to ensure that $\varepsilon_t \sim N(0,\ \sigma^2)$. The parameters ρ and b_j in (9) are estimated to fit the model and evaluate the null hypothesis of z being non-stationary. The null hypothesis is evaluated through the test-statistic t_ρ. If t_ρ is lower than the critical value from the Dickey-Fuller tables, z is stationary. The more negative the t_ρ obtained, the more stationary is the z [13].

The results of the ADF depend on the q selected for model (9). For this reason, the ADF of the set of z is performed for a range of $q_i \in [q_1\ q_2\ \dots\ q_v]$. From each test, the most stationary cointegrated residual and its associated cointegrated vector $\boldsymbol{\beta}_i$ are found. These cointegrated vectors $\boldsymbol{\beta}_i$, $i \in [1,\ v]$, are the ones that give the most stationary cointegrated residuals in the defined ranges of q and k. In this study, the range of values to be used for k is defined according to the procedure described in [14], which is also used for q, thus

$$\boldsymbol{q} = \boldsymbol{k} = \left[3,\ 12\left(\frac{N}{100}\right)^{1/4}\right]. \tag{10}$$

The procedure for finding the set of suitable cointegrated vectors $\boldsymbol{\beta}_i$ within these ranges is summarized in Fig. 1. If the training features are cointegrated, this procedure gives v suitable cointegrated vectors $\boldsymbol{\beta}_i$.

Once $\boldsymbol{\beta}_i$ has been found from the training features, (8) is used to compute the cointegrated residual of each of the features from the testing phase. Let $\mathcal{Z}_L$ and $\mathcal{Z}_U$ denote lower and upper limits, then the structural system in question is labelled as damaged if the testing phase residual is not confined to the interval $[\mathcal{Z}_L, \mathcal{Z}_L]$.

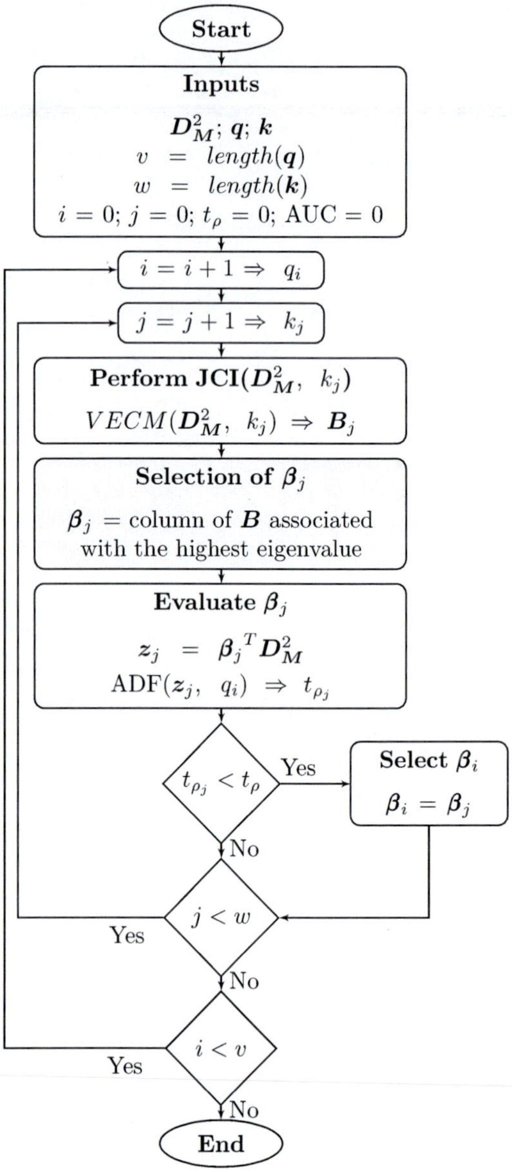

Fig. 1 Procedure to find suitable cointegrated vectors β_i from the range of lag values for k and q

3 Laboratory Experiment

The performances of the two VSDD schemes outlined in the methodology section are evaluated experimentally based on a cantilevered steel beam, which is exposed to temperature-based EVs.

3.1 Experimental Setup

The beam, which is depicted in Fig. 2a, is instrumented with nine accelerometers that sample with a frequency of 8192 Hz. In each observation/experimental trial, the accelerometers capture 15 s of the response induced by an impulse load applied to the bottom of the beam. The beam system is analyzed in its reference/undamaged state and in three damaged states, where, as indicated in Fig. 2b, masses of 1, 2.5, and 4.3 g are added in the free end of the beam. Temperature variability is introduced in all structural states by use of a heating cable, which is attached to the beam as seen in Fig. 2c. A thermostat controls the temperature sensed by a NTC thermistor fixed to the beam, and an infrared thermometer is used to measure the temperature.

(a) (b) (c)

Fig. 2 Cantilevered beam for the laboratory experiment. (**a**) Nine accelerometers mounted on the front of the beam. (**b**) Mass of 4.3 g. Added to the top of the beam to emulate damage. (**c**) Heating cable attached to the back of the beam

The temperature variability is shown in Fig. 3. Evidently, the variability differs for the different structural states, which has been chosen to test the efficiency of the VSDD schemes when some of the EVs are not accounted for in the baseline models.

3.2 Data Acquisition and Feature Extraction

From the 15 s period of available vibration data, an interval of 0.5 s (4096 sample points) is extracted and stored in $\boldsymbol{y}_{f,t} \in \mathbb{R}^{1x4096}$ for further analysis. Training and testing observation sets are obtained from the undamaged state of the beam, and a set of observations is obtained from each of the three damaged states. The features $\boldsymbol{x}_t$ for the MD-based scheme are computed from (3). For the cointegration-based scheme, we plug $\boldsymbol{c}_{f,t} \in \mathbb{R}^{4096}$ into (5) and attain the MDs for each sensor, which are gathered in $\boldsymbol{D}_M^2 \in \mathbb{R}^{9\times N}$. The suitable cointegrated vector(s) $\boldsymbol{\beta}_i$ are found from the training features by following the procedure outlined in Fig. 1. From each $\boldsymbol{\beta}_i$, the corresponding cointegrated residual z_i is computed by (8) for the training and testing observations.

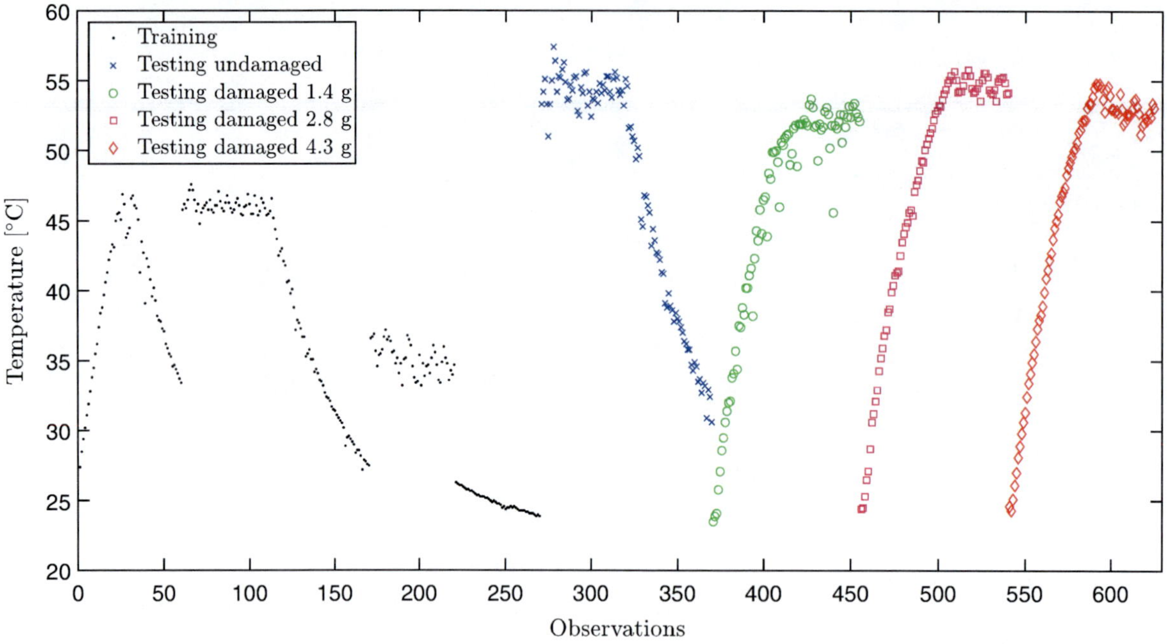

Fig. 3 Temperature field imposed on the beam

4 Results

The baseline models for the VSDD schemes are constructed from $N = 270$ training observations, while 100 observations from the undamaged state and 85 observations from each of the three damaged states are tested against the baseline models. Figure 4a shows the resulting $d^2_{M_t}$ from the MD-based scheme, and the results from the cointegration-based scheme with $q = 10$ and $k = 10$ are displayed in Fig. 4b. In the latter, the mitigation of the influence of the EVs can be clearly appreciated in the testing observations from the undamaged state. To quantify the performance of each scheme, FPRs, TPRs, and AUCs are computed from the damage indices. For the computation of FPRs and TPRs, a threshold of 1% is used in the MD-based scheme, while lower and upper limits of $\pm 1\%$ are used in the cointegration-based scheme. The computed values are summarized in Tables 1 and 2, where it can be seen that the cointegration-based scheme with lag values $q = 10$ and $k = 10$ provides the best performance.

5 Conclusion

The present paper offers a comparative study of two VSDD schemes for detecting mass perturbations in a laboratory beam system exposed to temperature-governed EVs. In particular, the performance of a recently proposed cointegration-based scheme is compared to that of a conventional MD-based outlier analysis scheme that does not account for the EVs. It is found that the cointegration-based scheme allows for a mitigation of the EVs, hence providing an improved damage detection performance compared to that provided by the conventional scheme. It must, however, be noted that the performance of the cointegration-based scheme hinges on a proper selection of the number of lags to include in the VECM and the cointegrated vector(s) to use in the computation of the cointegrated residual(s). To the authors' knowledge, a general procedure for this selection is not available.

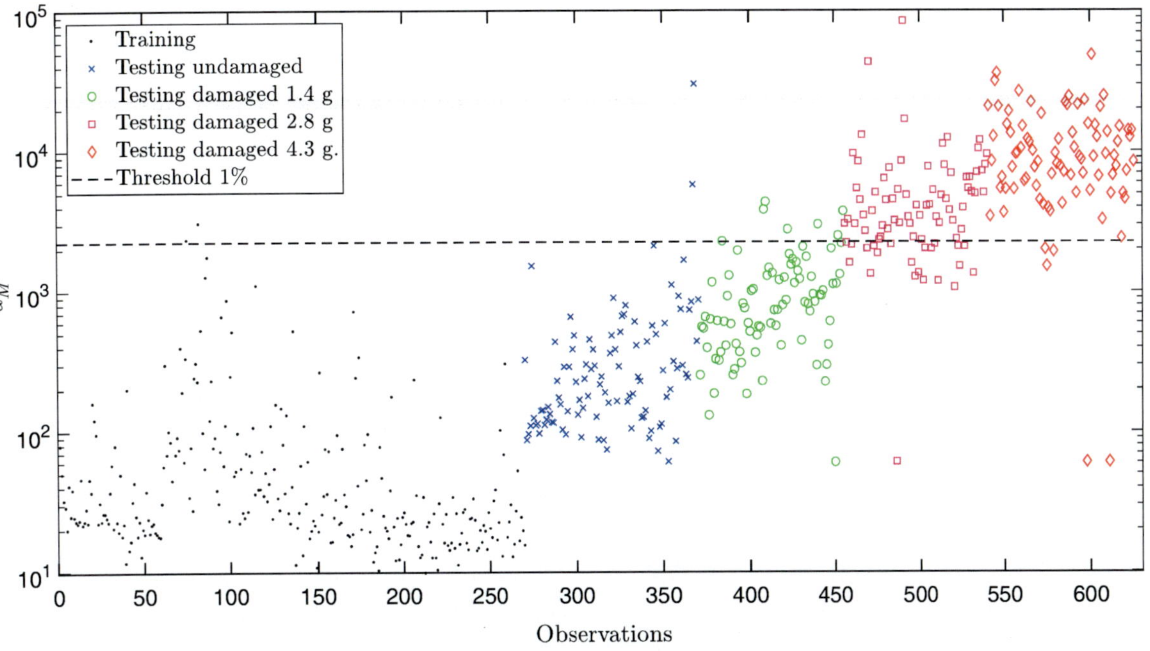

(a)

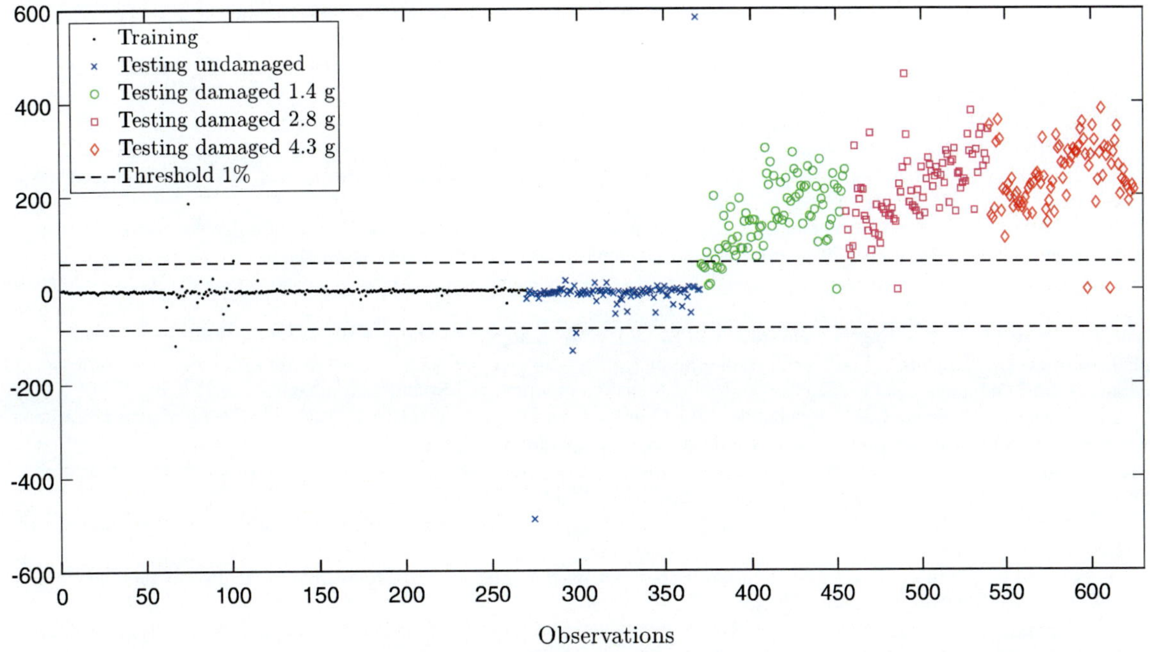

(b)

Fig. 4 Damage detection results. (**a**) MD-based results. (**b**) Cointegration-based results with $q = 10$ and $k = 10$

Table 1 Performance indicators of the MD-based scheme (FPR and TPR for 1% threshold)

AUC	FPR (%)	TPR (%)
0.9193	2.00	58.82

Table 2 Performance indicators of the cointegration-based scheme (FPR and TPR for limits of ±1%)

q	k	AUC	FPR (%)	TPR (%)
3	7	0.9282	3.00	29.80
4	12	0.9496	4.00	71.37
5	10	0.9456	4.00	75.46
6	11	0.9422	5.00	77.25
7	8	0.9161	2.00	28.62
8	8	0.9161	2.00	28.62
9	8	0.9161	2.00	28.62
10	10	0.9841	4.00	92.15
11	8	0.9161	2.00	28.62
12	11	0.9422	5.00	77.25
13	11	0.9422	5.00	77.25
14	12	0.9496	4.00	71.37
15	8	0.9161	2.00	28.62

References

1. Farrar, C.R., Worden, K.: Structural Health Monitoring: A Machine Learning Perspective, 1st edn. Wiley, Hoboken, NJ (2013)
2. Ulriksen, M.D.: Damage localization for structural health monitoring: an exploration of three new vibration-based schemes, Ph.D. thesis, Aalborg University, Aalborg (2018)
3. Wahab, M.A., Roeck, G.D.: Effect of temperature on dynamic system parameters of a highway bridge. Struct. Eng. Int. **7**(4), 266–270 (1997). https://doi.org/10.2749/101686697780494563
4. Peeters, B., De Roeck, G.: One-year monitoring of the z24-bridge: environmental effects versus damage events. Earthquake Eng. Struct. Dynam. **30**(2), 149–171 (2001). http://dx.doi.org/10.1002/1096-9845(200102)30:2<149::AID-EQE1>3.0.CO;2-Z
5. Augustyn, D., Smolka, U., Tygesen, U.T., Ulriksen, M.D., Sørensen, J.D.: Data-driven model updating of an offshore wind jacket substructure. Appl. Ocean Res. **104**, 102366 (2020). http://dx.doi.org/10.1016/j.apor.2020.102366
6. Qadri, B.A., Avendaño-Valencia, L.D., Hooghoudt, J.-O., Tcherniak, D., Ulriksen, M.D.: Removal of environmental and operational effects in damage detection: a comparative study with an operating wind turbine. Struct. Health Monit. (submitted)
7. Engle, R., Granger, C.: Co-integration and error correction: representation, estimation, and testing. Econometrica **55**(2), 251–276 (1987)
8. Cross, E.J., Worden, K., Chen, Q.: Cointegration: a novel approach for the removal of environmental trends in structural health monitoring data. Proc. R. Soc. A: Math. Phys. Eng. Sci. **467**(2133), 2712–2732 (2011). https://doi.org/10.1098/rspa.2011.0023
9. Cross, E.J., Worden, K., Sturgeon, R.J., Chen, Q.: A tutorial on cointegration for engineers—a tool for non-stationary time series analysis. In: Proceedings of the 10th International Conference on Recent Advances in Structural Dynamics, Southampton (2010), pp. 12–14
10. Qadri, B.A., Avendaño-Valencia, L.D., Hooghoudt, J.-O., Ulriksen, M.D.: A cointegration-based scheme for damage detection under environmental and operational variability. Struct. Health Monit. (in preparation)
11. Tcherniak, D., Mølgaard, L.L.: Active vibration-based structural health monitoring system for wind turbine blade: demonstration on an operating Vestas V27 wind turbine. Struct. Health Monit. **16**(5), 536–550 (2017)
12. Johansen, S.: Statistical analysis of cointegration vectors. J. Econ. Dynam. Control **12**(2), 231–254 (1988). https://doi.org/10.1016/0165-1889(88)90041-3
13. Cross, E.J., Worden, K., Chen, Q.: Cointegration: a novel approach for the removal of environmental trends in structural health monitoring data. Proc. R. Soc. A Math. Phys. Eng. Sci. **467**(2133), 2712–2732 (2011)
14. Dao, P.B., Staszewski, W.J., Klepka, A.: Stationarity-based approach for the selection of lag length in cointegration analysis used for structural damage detection. Comput.-Aided Civil Infrastr. Eng. **32**(2), 138–153 (2017)

Footbridge Vibrations and Modelling of Pedestrian Loads

Lars Pedersen and Christian Frier

Abstract Vibrations in footbridges can be annoying and hence it is useful already at the design stage to be able to predict levels of footbridge vibrations in order to ensure that serviceability-limit-state requirements will be fulfilled. For the studies of the paper footbridge vibrations are assumed brought about by pedestrians. Walking parameters such as load amplification factors, pacing frequency, pacing speed, and pedestrian weight determine the characteristics of the loading. By nature, these parameters are stochastic and hence the studies of this paper will handle some of the walking parameters as random variables. This has the effect that predictions of footbridge vibration levels end up being random variables. The paper will consider and examine how selected decisions related to setting up the calculation framework can influence the outcome of design stage predictions of footbridge vibration levels.

Keywords Footbridge vibrations · Walking loads · Walking parameters · Stochastic load models · Serviceability limit state

Nomenclature

a	Bridge acceleration
i	Integer
v	Pacing speed
L	Bridge length
α	Dynamic load factor
σ	Standard deviation
f_1	Bridge fundamental frequency
m_1	Bridge modal mass
t	Time
Q	Modal load
ζ_1	Bridge damping ratio
Θ	Phase
f_s	Step frequency
l_s	Step length
F	Walking load
W	Weight of pedestrian
μ	Mean value
Φ	Mode shape

L. Pedersen (✉) · C. Frier
Department of Civil Engineering, Aalborg University, Aalborg Ø, Denmark
e-mail: lp@civil.aau.dk; cf@civil.aau.dk

K. Grimmelsman (ed.), *Dynamics of Civil Structures, Volume 2*, Conference Proceedings of the Society for Experimental Mechanics Series, https://doi.org/10.1007/978-3-030-77143-0_7

1 Introduction

A footbridge may be flexible and is potentially prone to vibrate due to human excitation. The dynamic forces caused by pedestrians might not cause ultimate-limit-state issues, but the serviceability limit state may be of concern.

A well-known and critical scenario was that occurring on the Millennium Bridge in London [1]. Here, both vertical and horizontal motion caused by pedestrians showed to be a problem. The problematic conditions often relate to coincidence between bridge natural frequencies and frequencies of human in motion. In this case excessive structural vibrations may occur and if not properly designed, the bridge may be unfit for its intended use.

In this paper focus is on the vertical action of pedestrians. The action has for decades been modelled as a deterministic load [2–4]. More recently, in for instance [5–8], the stochastic nature of the action has been considered and modelled.

Hence, this paper will adapt the stochastic line of thinking, hereby modelling central walking parameters as random variables. Doing so recognises that parameters such as pedestrian step frequency, step length and dynamic load factors are in fact stochastic properties by nature.

A comprehensive line-up of the probabilistic scenario is introduced in [9]. The present paper takes off-set in the general approach, but also considers, adapts and evaluates simplifications along the way.

For all investigations off-set is taken in artificial footbridges. They will be assumed pin-supported, and focus will be on the vertical bridge response at midspan in the form of accelerations. As a result of the fact that the loading is stochastic, the response will be stochastic too, and the output of calculations will be the acceleration quantiles. These are derived combining Newmark time integration with Monte Carlo simulations.

The paper aims at exploring how sensitive the stochastic nature of footbridge vibration is to decisions related to defining dynamic load factors and how sensitive it is in regard to assumptions made for the stochastic nature of step frequencies of the pedestrians.

Section 2 describes basic model assumptions.

Section 3 describes the different study angles of the paper, and Sect. 4 summarises conclusions.

2 Modelling of Walking Loads

The walking load models considered in this paper rely on a modal load assumption in which the modal load $Q(t)$ acting on the footbridge and generated by the pedestrian is derived using Eq. (1).

$$Q(t) = \Phi(t)F(t) \tag{1}$$

$F(t)$ represents the vertical force imparted at the position of the pedestrian while crossing the bridge, and $\Phi(t)$ is the mode shape function. Only the first bending mode is considered to dominate the response, and hence this will be taken as a half-sine sinusoidal. It will depend on the pacing speed v of the pedestrian, and in consequence hereof on the step frequency f_s and step length l_s of the pedestrian as a result of the relationship shown in Eq. (2).

$$v = f_s l_s \tag{2}$$

The mode shape function is calculated using Eq. (3).

$$\Phi(t) = \sin(\pi v t / L) \tag{3}$$

This load model, $F(t)$, is the model introduced in [9]. The mathematical expression for $F(t)$ is seen in Eqs. (4–6):

$$F(t) = \sum_{i=1}^{5} F_i(t) + \sum_{i=1}^{5} F_i^S(t) \tag{4}$$

$$F_i(t) = W\alpha_i \sum_{\overline{f}_j=i-0.25}^{i+0.25} \overline{\alpha}_i(\overline{f}_j)\cos\left(2\pi\overline{f}_j f_s t + \theta(\overline{f}_j)\right) \tag{5}$$

$$F_i^S(t) = W\alpha_i^S \sum_{\overline{f}_j^S=i-0.75}^{i-0.25} \overline{\alpha}_i^S\left(\overline{f}_j^S\right)\cos\left(2\pi\overline{f}_j^S f_s t + \theta\left(\overline{f}_j^S\right)\right) \tag{6}$$

Reference is made to [9], for a detailed description.

Here it suffices to mention that W represents the static weight of the pedestrian. Furthermore, that f_s represents the step frequency.

The model consists of main load harmonics (Eq. 5) and subharmonics (Eq. 6). The latter due to the fact that "the fundamental period of the force time history is equal to the time required to make two successive steps, rather than one" [9].

A governing parameter for the loading is the dynamic load factors, α_i, $(i = 1, 2, \ldots, 5)$.

For the main harmonic, α_1 (the first load harmonic), the following mean value, μ, and a standard variation, σ, is assumed.

$$\mu = -0.2649 f_S^3 + 1.3206 f_S^2 - 1.7597 f_S + 0.7613; \quad \sigma = 0.16\mu \tag{7}$$

As would appear, the dynamic load factor is modelled as a random variable, and the distribution is assumed Gaussian.

Table 1 defines the assumptions made for the dynamic load factors, for $\alpha_i (i = 2, 3, 4, 5)$, and the corresponding mean values (μ) and standard deviations (σ).

The subharmonic load factors α_i^S are derived from the main harmonic load factor, α_1, in the way described in [9].

Having set up the load, Newmark time-integration allows for computing bridge response and Monte Carlo simulations for establishing a statistical basic for the response. From this, acceleration quantiles can be derived, such as the acceleration quantile a_{95}, which will be the parameter in focus in this paper for describing the acceleration level of the considered footbridges.

For each bridge 100.000 simulations were conducted.

3 Studies of This Paper

3.1 Impact of Decisions Related to Modelling the Load Amplification Factor

The purpose of this study is to explore how different decisions related to modelling the dynamic load factor influence estimates of the stochastic bridge response in the form of the acceleration quantile a_{95}.

A simplified load model is assumed, namely one that only considers the first harmonic, α_1. Leakage of energy around this load harmonic is also disregarded.

The following assumptions were made as regards the walking parameters, see Table 2.

Hence, it is assumed that step length and step frequency are random variables, and Gaussian distributions are assumed to apply.

By this approach, the load amplification factor (Eq. 7) would vary depending on the outcome of f_s in simulations from one pedestrian crossing to the next.

Table 1 Mean values and standard deviations

–	α_2	α_3	α_4	α_5
μ	0.07	0.05	0.05	0.03
σ	0.030	0.020	0.020	0.015

Table 2 Mean values and standard deviations

–	μ	σ	Reference
W	750 N	0 N	[9]
l_s	0.71 m	0.071 m	[9]
f_s	1.87 Hz	0.186 Hz	[9]

An alternative assumption for the calculations could be to assume that the load amplification factor assumes a constant value, namely the value that can be calculated by assuming the mean value of the step frequency f_s for the calculations of the dynamic load factor (in Eq. 7).

This was done for the SDOF footbridges listed in Table 3.

The combination of values of f_1, m_1, and L is believed to be fairly realistic for SDOF pin-supported footbridges, as m_1 and L drop with increase in f_1.

In terms of the deviations between outcomes of calculations of a_{95}, obtained by the two approaches, Table 4 shows the results. The difference between results is normalised to the result obtained assuming that the dynamic load factor would vary from one bridge crossing to the next.

The results signify that it may not be totally off to employ a simplified approach for computations, when it comes to settling on the dynamic load factor. Even though the simplified approach somewhat violates the stochastic nature of the problem, fairly reasonable results are obtained.

It is underlined that the investigations presented here do not reflect the uncertainty of the stochastic nature of the dynamic load factor. For such information see for instance [8].

3.2 *Impact of Decisions Related to Modelling the Step Frequency*

To this end of investigation different sets of assumptions related to modelling the stochastic nature of step frequency are considered.

The step frequency will be modelled as a random variable, but assumptions related to mean value and standard deviation need to be made.

For the investigations of this paper, the assumptions listed in Table 5 are considered.

The models represent different proposals that can be found in literature [8].

Here, they are considered as input data for computing a_{95} for different footbridges.

For the study, the bridges tabulated in Table 6 are considered.

The dynamic characteristics for the bridges do not exactly correspond to those in the previous study, but this is not of primary importance.

Table 3 Modal properties of bridges and bridge lengths (L)

		Bridge		
Property	Unit	A	B	C
f_1	Hz	1.6	1.9	2.2
ζ_1	%	0.5	0.5	0.5
m_1	10^3 kg	61.5	44.0	32.5
L	m	54.0	45.0	39.0

Table 4 Deviations in terms of estimates of a_{95}

		Bridge		
Deviation	Unit	A	B	C
a_{95}	%	+10	1	−20

Table 5 Mean values and standard deviations

Model	Unit	μ	σ
I	Hz	1.87	0.186
II	Hz	1.99	0.173
III	Hz	2.20	0.300

Table 6 Modal properties of bridges and bridge lengths (L)

		Bridge		
Property	Unit	A	B	C
f_1	Hz	1.85	2.00	2.20
ζ_1	%	0.5	0.5	0.5
m_1	10^3 kg	46.2	39.5	32.6
L	m	46.5	43.0	39.1

Table 7 Acceleration quantile a_{95}

	Model for step frequency		
Bridge	I (m/s^2)	II (m/s^2)	III (m/s^2)
A	0.3093	0.2879	0.1750
B	0.3904	0.4256	0.3442
C	0.2743	0.4715	0.5033

In terms of bridge response, and focusing on the response characteristic a_{95}, Table 7 summarises the results computed for footbridge A, B, and C. Values of a_{95} are provided for the three different step frequency models.

For bridge A there is a maximum difference of (0.3093–0.1750=) 0.1343 m/s^2 between results obtained for the three stochastic models for step frequency. If this value is normalised by the minimum number 0.1750 m/s^2, one obtains a 77% difference in results for bridge A.

Doing the same calculations for bridge B and C results in 24% and 83% differences, respectively.

Hence, the choice of step frequency model assumed for computing a_{95} has a relative high impact on the result.

4 Conclusion and Discussion

In the paper the influence of decisions as regards settling on a framework for pedestrian load models for evaluating footbridge response at the design stage was examined. Focus was on estimation of the footbridge acceleration response occurring at midspan of footbridges. The acceleration quantile a_{95} (the acceleration level exceeded in 5% of the pedestrian crossings) was chosen for investigation.

For the investigations, different artificial SDOF and pin-supported bridges were considered so as to widen the perspective of conclusions.

One issue addressed was on ways for choosing the dynamic load amplification factor for a computational prediction of footbridge response. Another issue was on choosing parameters of a stochastic model for the step frequency of pedestrians for entering into the calculations.

Both choices might potentially affect the outcome of the predicted stochastic nature of bridge response and hence serviceability-limit-state evaluations for footbridges.

As for the dynamic load factor, different methods for extracting the main governing load amplification factor were examined. It turned out that a simplified technique not fully in accordance with the stochastic nature of the pedestrian traffic provided fairly reasonable results (errors in predictions of a_{95} of maximum 20% for the investigated bridges).

Whereas the investigations in terms of the dynamic load factor focused on a technique for simpler processing of data, the investigations in terms of choosing parameters for a stochastic model for step frequency directly relate to actual uncertainties.

Solutions to this challenge are not provided here but it is interesting to notice that the calculations of this paper suggest up to 83% deviations in estimates of a_{95} depending on which bridge is considered and which input parameters are chosen for modelling the stochastic nature of step frequencies.

References

1. Dallard, P., Fitzpatrick, A.J., Flint, A., Le Bourva, S., Low, A., Ridsdill-Smith, R.M., Wilford, M.: The London millennium bridge. Struct. Eng. **79**, 17–33 (2001)
2. Ellis, B.R.: On the response of long-span floors to walking loads generated by individuals and crowds. Struct. Eng. **78**, 1–25 (2000)
3. Bachmann, H., Ammann, W.: Vibrations in Structures – Induced by Man and Machines. IABSE Structural Engineering Documents 3e, Zürich, Switzerland (1987)
4. Rainer, J.H., Pernica, G., Allen, D.E.: Dynamic loading and response of footbridges. Can. J. Civ. Eng. **15**, 66–78 (1998)
5. Matsumoto, Y., Nishioka, T., Shiojiri, H., Matsuzaki, K.: Dynamic design of footbridges. In: IABSE Proceedings, No. P-17/78, pp. 1–15 (1978)
6. Živanovic, S.: Probability-based estimation of vibration for pedestrian structures due to walking. PhD Thesis, Department of Civil and Structural Engineering, University of Sheffield, UK (2006)
7. Kerr, S.C., Bishop, N.W.M.: Human induced loading on flexible staircases. Eng. Struct. **23**, 37–45 (2001)
8. Pedersen, L., Frier, C.: Sensitivity of footbridge vibrations to stochastic walking parameters. J. Sound Vibr. (2009). https://doi.org/10.1016/j.jsv.2009.12.022
9. Živanovic, S., Pavic, A., Reynolds, P.: Probability-based prediction of multi-mode vibration response to walking excitation. Eng. Struct. **29**, 942–954 (2007). https://doi.org/10.1016/j.engstruct.2006.07.004

Multi-LSTM-Based Framework for Ambient Intelligence

Nur Sila Gulgec, Martin Takáč, and Shamim N. Pakzad

Abstract Bridge structures experience significant vibrations, repeated stress cycles, and deterioration during their life cycles. Thus, it is essential to accurately establish life cycle assessment which often needs collecting strain responses. Conventional way of measuring strain measurements is laboriously ineffective and expensive as more spatial information is desired. In the presented approach, inexpensively collected acceleration responses by using wireless sensor networks or mobile sensing are utilized to predict strain information and decrease the installation expenses of sensors systems, as well as enable crowdsourcing potential of the mobile sensing. The study employs a deep learning framework composed of fully connected (FC) and multiple Long Short-Term Memory (LSTM) layers to predict strain time series from the acceleration responses. The deep architecture achieves to learn the relationship between input and output by exploiting the temporal dependencies of them. In the evaluation of the method, acceleration data collected from a steel bridge is utilized to predict the strain time series. The learned architecture is tested on an acceleration time series that the structure has never experienced. The initial findings show that this study holds great potential to perform sensing of the structures with affordable and intelligent sensing systems.

Keywords Structural health monitoring · Deep learning · Sensing · Sensor networks · Mobile sensing · Long short-term memory

1 Introduction

The life cycle assessment requires designing computationally affordable performance indicators for both the components of the structure and its overall system. Building finite element (FE) models of structures' systems and components help engineers to provide more accurate performance assessment. Traditionally, structural health monitoring (SHM) data is used to update FE models and lifetime reliability of structures [1]. In reliability analysis, the primary monitored response quantity measured by SHM is strain [2–4]. The life cycle performance is updated by the collected SHM data and the maintenance optimization is performed to assess structures' life cycle and update its service life.

Traditional inspection methods collect strain measurements by using strain; nevertheless, large-scale and spatially dense installation of wired strain gauges is expensive and laboriously impractical [5]. To address these limitations, the integration of information from inexpensive data sources is necessary [6, 7]. Acceleration data can be collected relatively inexpensively by means of fixed acceleration sensors, wireless sensor networks, and mobile sensing. Mobile sensing is an alternative paradigm to collect comprehensive spatial information by using a few sensors. Mobile sensors have low setup costs compared to the traditional stationary sensor networks and they do not require to be dedicated to any particular structure [8]. Most importantly, when the ubiquity of smartphones with the internet of things (IoT) connectivity is considered, cars+smart phones can be treated as large-scale sensor networks which can contribute to the life cycle assessment of the structures on a daily basis [9].

N. S. Gulgec (✉)
Thornton Tomasetti, San Francisco, CA, USA

M. Takáč
Department of Industrial and Systems Engineering, Lehigh University, Bethlehem, PA, USA
e-mail: mat614@lehigh.edu

S. N. Pakzad
Department of Civil and Environmental Engineering, Lehigh University, Bethlehem, PA, USA
e-mail: snp208@lehigh.edu

K. Grimmelsman (ed.), *Dynamics of Civil Structures, Volume 2*, Conference Proceedings of the Society for Experimental Mechanics Series, https://doi.org/10.1007/978-3-030-77143-0_8

The authors propose a Long Short-Term Memory (LSTM) based framework to estimate the stress or strain responses from the acceleration information. The proposed technique aims to build a relationship between acceleration and strain for a few selected locations under the loading conditions included in the training dataset. In the testing phase, acceleration data from desired locations (i.e., the number of desired locations is greater than the selected ones) are fed into the architecture to predict strain responses. This approach is validated on ambient data collected from a bridge over 2 months.

The rest of the paper is organized as follows. First, related work and background information on deep learning are provided in Sects. 2 and 3, respectively; then, the proposed approach is described in Sect. 4. In Sects. 5 and 6, the experimental setup and the proposed network architecture are introduced. The main findings of this study are discussed in Sect. 7. Conclusions and future work are presented in Sect. 8.

2 Related Work

Several studies have investigated how expensive sensor measurements can be obtained by using inexpensively collected vibration data. Determining displacement responses from accelerometers has been commonly studied due to its challenges such as sensor and installation cost, and the inaccessibility of a reference point for full-scale structures [10–12]. The quantitative results showed that displacements could be obtained by utilizing acceleration responses.

The acceleration data has also been utilized to estimate strain responses at unmeasured locations when direct measurement is not possible; this approach is called virtual sensing. It is possible to group the virtual response estimation techniques into three main groups, namely, Kalman filtering [13, 14], a joint input-state estimation [15], and modal expansion [16, 17]. It is also discussed that multi-sensor data has the potential to improve the performance of response estimation [14]. Among the combination of input variables measuring accelerations, strains, and tilts to see the effect on the performance of strain estimations, the highest performance is found to be in the fusion of acceleration and tilt.

Previous research achieved promising results showing that strain response estimation is possible with acceleration data. Nevertheless, they suffer from several limitations. The mentioned techniques are dependent on the accuracy of the finite element models. Moreover, the majority of these approaches focus on zero-mean stationary processes and assume the unknown system input as zero-mean white noise which restricts the application domain. The objective of this study is to overcome these limitations by proposing a data-driven, deep learning-based approach which can be generalizable to different kind of vibrations and overcome the problem of learning long-time dependencies.

3 Background

Creating entirely accurate and detailed structural models is a very challenging task due to the complexity of structures. Thus, data-driven approaches become prominent for SHM applications where a surrogate model, constructed using collected structural responses, is substituted for a real model [18]. With today's advanced technology enabling crowdsourcing, the abundance of data is an opportunity to extract information about infrastructure condition throughout their entire life.

Utilizing deep neural networks (DNNs) is an ideal solution for taking advantage of this massive data [19]. The output of DNNs is formed by nonlinear mapping of the input data; therefore, their performance improves when more data is available. DNNs consist of multiple linear layers and nonlinear transformations, where each layer is trained to extract some relevant information. After this training step, the learned DNN architecture provides informed decisions based on the data-driven predictions.

The implementation of DNNs varies greatly depending on the application domain. LSTM networks proposed by Hochreiter and Schmidhuber [20] are widely used to exhibit temporal dynamic behavior for a time sequence. LSTMs are a type of recurrent neural networks (RNNs) where the present decisions (at time step t) are affected by the recent past (at time step $t - 1$). LSTMs comprise a chain of recurrent nodes forming memory cells. These cells remember the information and provide control over which information to let through and which information to forget with the help of input gate, forget gate, and output gate. The Refs. [20, 21] provide more detailed information and equations about the architecture of LSTMs.

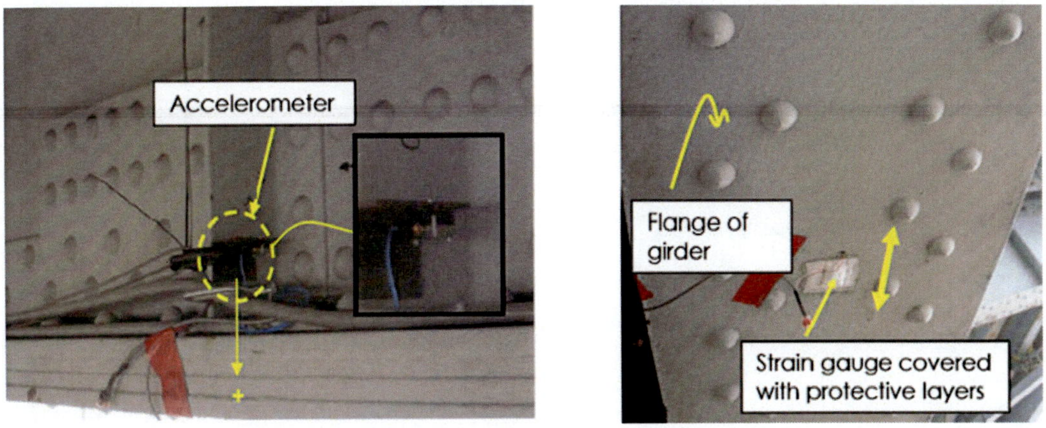

Fig. 1 The installed accelerometer and strain gauge in Span 3

4 Overview of the Proposed Approach

The proposed technique comprises two phases: training and testing. The training phase aims to build a relationship between acceleration and strain for a few selected locations under the loading conditions included in the training dataset. Such relationship is found with a deep learning architecture which is trained to minimize the error between the predicted and the actual strain responses. In the testing phase, acceleration data from desired locations (i.e., desired locations are greater than the selected ones) are fed into the architecture to predict strain responses.

In the framework, the acceleration responses can be collected by using fixed/wireless acceleration sensors or mobile sensing. Mobile sensors simultaneously measure vibration data in time while scanning over a large set of points in space which creates a problematic data arrangement. Since collected data is mixed in time and space, spatial discontinuities are often observed. If the acceleration information is collected by mobile sensing, the collected mobile data should be converted to regular acceleration time histories. There are several techniques available in the literature to provide such conversion [7, 9]. The converted acceleration information is then utilized for training the LSTM-based network architecture.

5 Experimental Setup and Data Preparation

The proposed approach is validated on a bridge consisting of the main girders, floor and sub-floor beams, lateral bracing system, stringers, and an orthotropic deck. Long-term monitoring of five approach spans was conducted where the approach spans are leading to the main suspension span of a bridge. Spans 1–4 are 58 m long two-girder simple spans with and orthotropic deck; whereas, Span 5 is a curved girder simple span with a horizontal length of approximately 50 m. It is worthwhile to mention that all spans have a similar superstructure.

Instrumented locations were monitored by accelerometers and strain gauges for approximately just over 2 months. The installed accelerometers and strain gauges in Span 3 are visualized in Fig. 1. Uniaxial accelerometers (i.e., model of 3701G3FA3G manufactured by PCB Piezotronics, Inc.) were installed to record the vertical accelerations of the girders at midspan. The model of accelerometers was selected to be able to record low-amplitude, low-frequency accelerations. The strains were measured by five strain gauges attached next to the accelerometers. The strain gauges recorded data only when the measured stress response exceeds predefined triggers.

The responses were collected on different days for different durations. Therefore, the longer time series are divided into subsequences with the size of the smallest sequence. Subsequences are normalized between [−1,1]. The example visualization of the collected sequence from Span 3 before the normalization step is visualized in Fig. 2. As can be observed from the figure, learning the relationship between acceleration and strain is a challenging task. For instance, when the acceleration levels of Span 3 are considered, it can be observed the increase in acceleration does not cause an increase in the strains.

The Span 3 and 4 acceleration and strain recordings are used to generate the training and validation dataset. Seventy percent and 15% of the subsequences are included in training and validation, respectively. While preparing the testing dataset, only acceleration data (i.e., 15% of the acceleration subsequences) from the remaining spans are utilized.

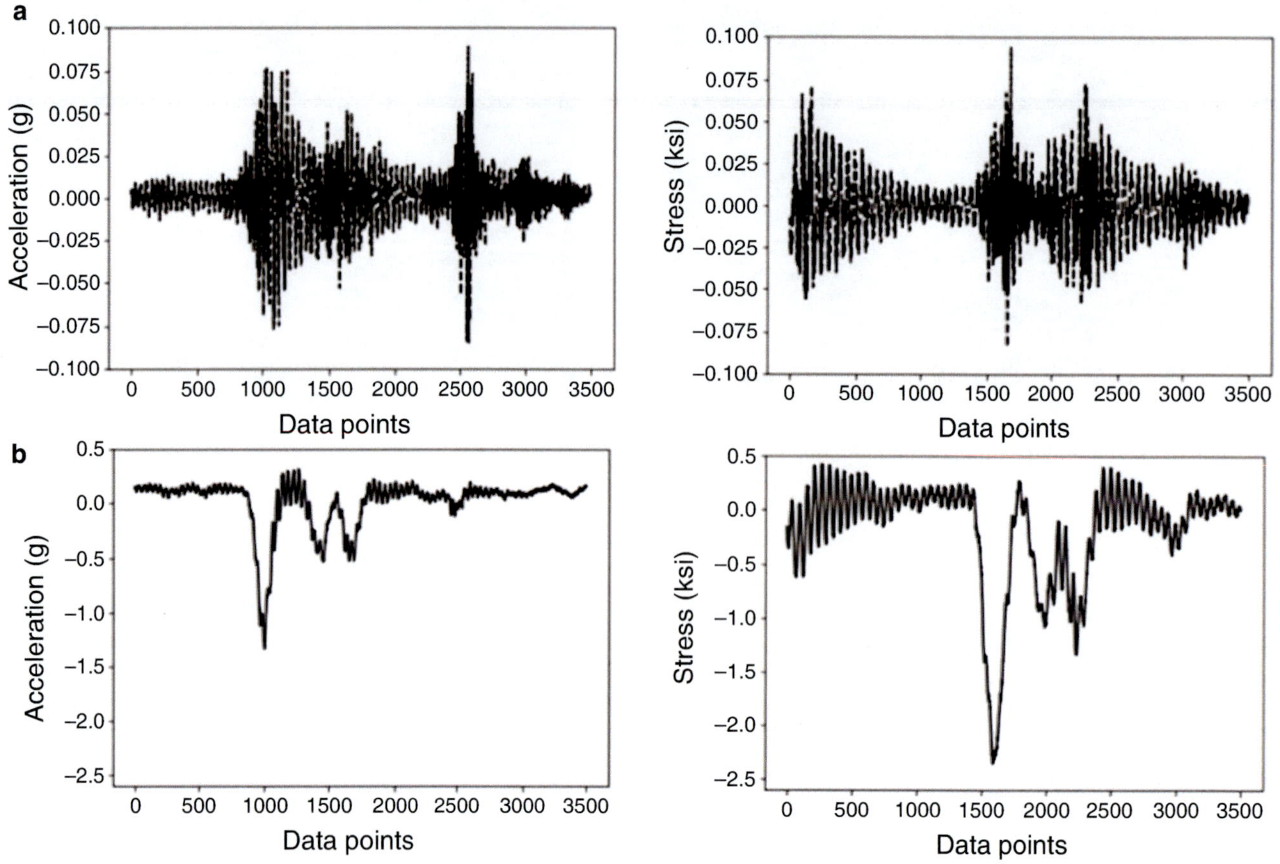

Fig. 2 An example sequence from (**a**) Span 3 and (**b**) Span 4

6 Proposed Network Architecture and Training

There are various ways to construct deep neural networks by utilizing the sequence of LSTM and fully connected (FC) layers. The accuracy and robustness of these networks are highly dependent on the hyperparameter selection [22]. The hyperparameter finding method introduced in [23] is implemented to find DNN architecture with good configuration. A variety of networks are built with the following hyperparameters: the number of LSTM layers [min: 1-max: 5] and FC layers [min: 1-max: 3], the number of hidden states [min: 21-max: 29], and the number of hidden layer sizes [min: 21-max: 29]. All networks are run for one epoch, then the 10% of the networks with the worst validation score is removed. The procedure is repeated with the remained ones until the best five networks are in the pool.

The DNN topology found by this search is visualized in Fig. 3. The proposed architecture receives the sequences of acceleration data in the form of [batch size, sequence length, span number]. At time t, the input sequences are used to feel two LSTM memory cells with a size of [256]. The output of the final LSTM cell is passed through the three FC layers size of [256]. The resultant of the FC layer is used to predict the strain sequences in a similar form of [batch size, sequence length, span number]. The activation function of tan$h$() and the dropout rate of 0.5 are used in the FC layers.

The network is trained to minimize the mean squared error between the predicted and the true strain time histories. ADAM optimizer with batch size of 128 is utilized during training [24]. Short-term dependencies exist in strain time histories; therefore, truncated backpropagation through time (BPTT) method is adopted [25].

7 Discussion of Results

After training the network by using the acceleration-strain pairs from Span 3 and 4, the model is tested on Span 1, 2, and 5. It is worthwhile to say that the distance between Span 2–3 and Span 4–5 is approximately 116 m and 404 m, respectively. The

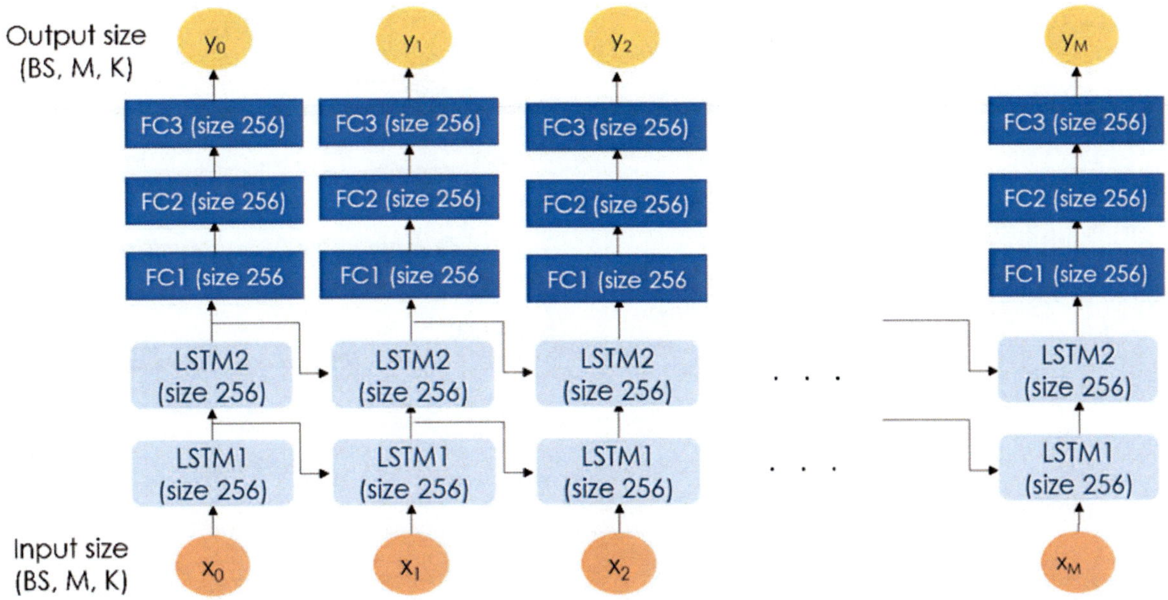

Fig. 3 The proposed network topology

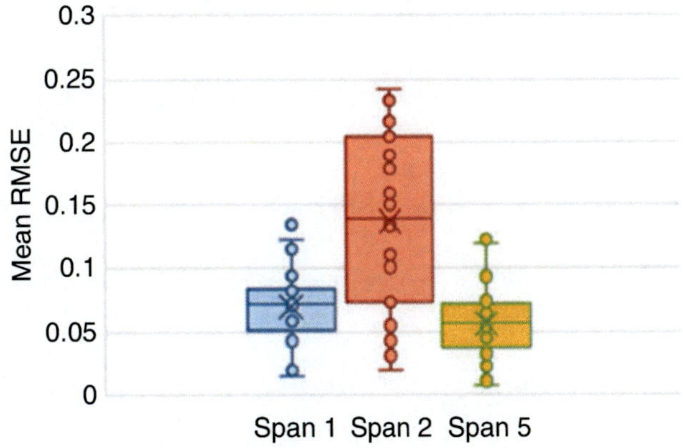

Fig. 4 The boxplot of RMSE values for the testing spans

boxplot of root mean squared (RMSE) error of the normalized strains are visualized in Fig. 4. Figure shows that the error is varying between 6% and 12%. Considering the difficulty of the task, these results are promising.

Several example predictions are plotted in Fig. 5 to investigate the sources of errors. In Fig. 5a, the error is associated with the high amplitude of the acceleration at the beginning of the subsequence. The initial values of the acceleration subsequences are typically closer to zero, unlike the case displayed in Fig. 5a. The second source of error is due to the initialization of LSTM cells which is shown in Fig. 5b. The algorithm needs several iterations to start predicting accurate results. The final source of error is about the bridge having an orthotropic deck. Orthotropic decks are typically light in weight which causes varying interaction between vehicles crossing the bridge and the deck. This interaction often depends on the vehicle type and the suspension system of the vehicle. If there is an excessive vibration at the time of the monitoring, it can result in a higher dynamic amplification factor of the girder than what is observed during the data collection.

8 Conclusions and Future Work

In this paper, deep learning-based framework is introduced to obtain strain or stress time histories by collecting acceleration data from inexpensive sensors. The proposed architecture consists of multiple LSTM and FC layers which help to discover temporal dependency and the complex relationship between acceleration and strain sequences. According to the findings

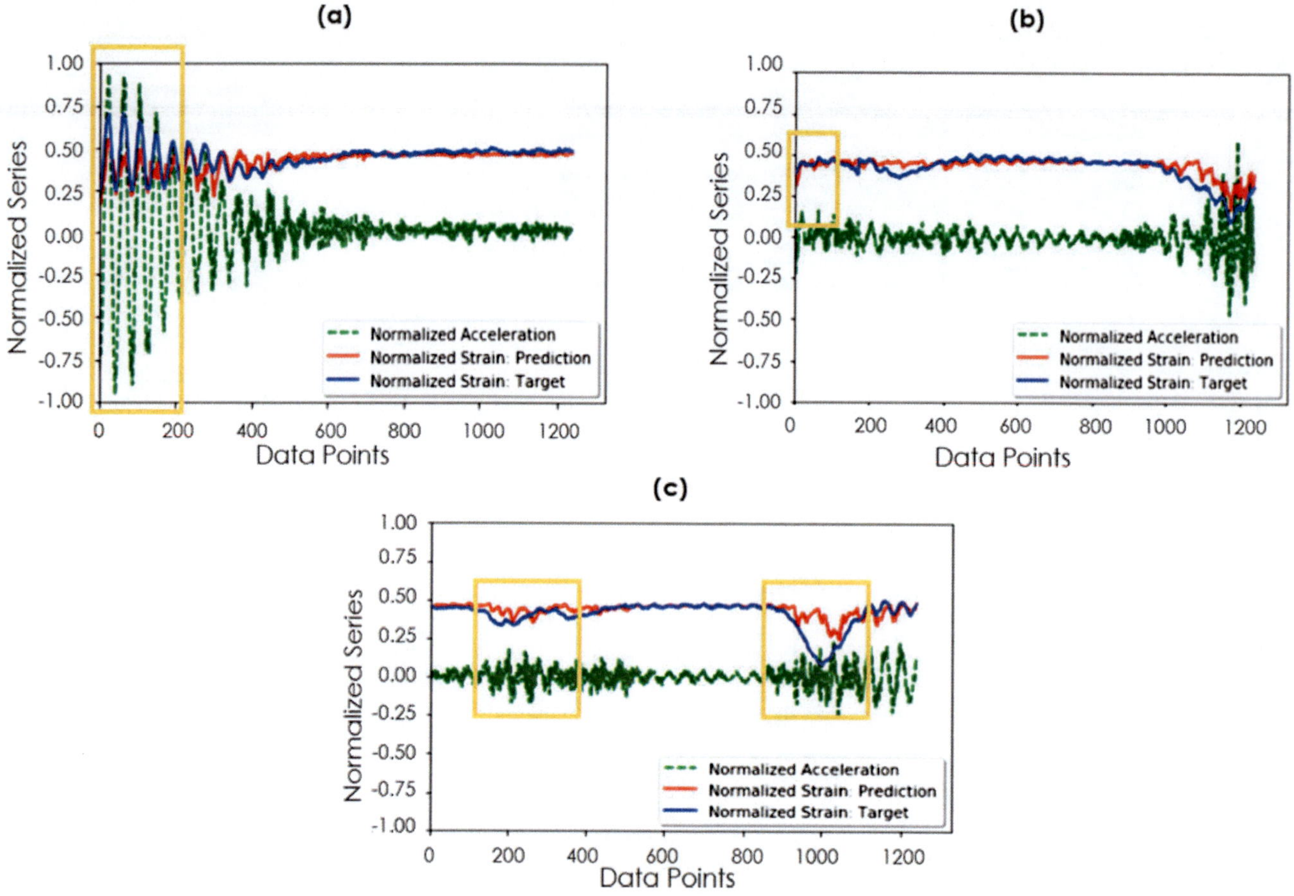

Fig. 5 Example predictions from Span 2 with different source of errors highlighted with yellow color

of the methodology, accurate strain prediction from acceleration data is challenging, yet promising for many potential applications, including life cycle assessment.

The future work aims to extend this research by generalizing the algorithm to different kind of bridge structures by considering both temporal and spatial dependencies of mobile sensing data. The proposed approach excludes the cases where desired measurement locations have high-stress concentrations due to damages. In future work, this limitation will also be focused and targeted.

Acknowledgments Research funding is partially provided by the National Science Foundation through Grant No. CMMI-1351537 by Hazard Mitigation and Structural Engineering program, by a grant from the Center for Integrated Asset Management for Multimodal Transportation Infrastructure Systems (CIAMTIS) and by a grant from the Commonwealth of Pennsylvania, Department of Community and Economic Development, through the Pennsylvania Infrastructure Technology Alliance (PITA). Martin Takáč was supported by National Science Foundation grants CCF-1618717, CMMI-1663256, and NSF:CCF:1740796.

References

1. Okasha, N.M., Frangopol, D.M., Orcesi, A.D.: Automated finite element updating using strain data for the lifetime reliability assessment of bridges. Reliab. Eng. Syst. Saf. **99**, 139–150 (2012)
2. Frangopol, D.M., Strauss, A., Kim, S.: Bridge reliability assessment based on monitoring. J. Bridg. Eng. **13**(3), 258–270 (2008)
3. Liu, M., Frangopol, D.M., Kim, S.: Bridge safety evaluation based on monitored live load effects. J. Bridg. Eng. **14**(4), 257–269 (2009)
4. Catbas, F.N., Susoy, M., Frangopol, D.M.: Structural health monitoring and reliability estimation: long span truss bridge application with environmental monitoring data. Eng. Struct. **30**(9), 2347–2359 (2008)
5. Cerda, F., Garrett, J., Bielak, J., Bhagavatula, R., Kovacevic, J.: Exploring indirect vehicle-bridge interaction for bridge SHM. In: Proc. of 5th Intl. Conf. on Bridge Maintenance, Safety & Management, Phil., pp. 696–702 (2010)

6. Gulgec, N.S., Takáč, M., Pakzad, S.N.: Innovative sensing by using deep learning framework. In: Dynamics of Civil Structures, vol. 2, pp. 293–300. Springer, New York (2019). https://doi.org/10.1007/978-3-319-74,421-6_39
7. Gulgec, N.S., Takáč, M., Pakzad, S.N.: Structural sensing with deep learning: strain estimation from acceleration data for fatigue assessment. In: Computer-Aided Civil and Infrastructure Engineering (2020)
8. Eshkevari, S., Pakzad, S.: Bridge structural identification using moving vehicle acceleration measurements. In: Dynamics of Civil Structures, vol. 2, pp. 251–261. Springer, New York (2019)
9. Eshkevari, S.S., Pakzad, S.N.: Signal reconstruction from mobile sensors network using matrix completion approach. In: Topics in Modal Analysis & Testing, vol. 8, pp. 61–75. Springer, Cham (2020)
10. Gindy, M., Vaccaro, R., Nassif, H., Velde, J.: A state-space approach for deriving bridge displacement from acceleration. Comput Aided Civ Infrastruct Eng. **23**(4), 281–290 (2008)
11. Sekiya, H., Kimura, K., Miki, C.: Technique for determining bridge displacement response using MEMS accelerometers. Sensors. **16**(2), 257 (2016)
12. Park, J.W., Sim, S.H., Jung, H.J.: Displacement estimation using multimetric data fusion. IEEE/ASME Trans. Mechatronics. **18**(6), 1675–1682 (2013)
13. Papadimitriou, C., Fritzen, C.P., Kraemer, P., Ntotsios, E.: Fatigue predictions in entire body of metallic structures from a limited number of vibration sensors using Kalman filtering. Struct. Control. Health Monit. **18**(5), 554–573 (2011)
14. Palanisamy, R., Cho, S., Kim, H., Sim, S.H.: Experimental validation of Kalman filter-based strain estimation in structures subjected to non-zero mean input. Smart Struct. Syst. **15**(2), 489–503 (2015)
15. Van der Male, P., Lourens, E.: Estimation of accumulated fatigue damage in lattice support structures from operational vibrations. EWEA Offshore. (2015)
16. Hjelm, H.P., Brincker, R., Graugaard-Jensen, J., Munch, K.: Determination of stress histories in structures by natural input modal analysis. In: Proceedings of 23rd Conference and Exposition on Structural Dynamics (IMACXXIII) (2005)
17. Maes, K., Iliopoulos, A., Weijtjens, W., Devriendt, C., Lombaert, G.: Dynamic strain estimation for fatigue assessment of an offshore monopile wind turbine using filtering and modal expansion algorithms. Mech. Syst. Signal Process. **76**, 592–611 (2016)
18. Sen, D., Nagarajaiah, S.: Data-driven approach to structural health monitoring using statistical learning algorithms. In: Mechatronics for Cultural Heritage and Civil Eng, pp. 295–305. Springer, Cham (2018)
19. Najafabadi, M.M., Villanustre, F., Khoshgoftaar, T.M., Seliya, N., Wald, R., Muharemagic, E.: Deep learning applications and challenges in big data analytics. J. Big Data. **2**(1), 1 (2015)
20. Hochreiter, S., Schmidhuber, J.: Long short-term memory. Neural Comput. **9**(8), 1735–1780 (1997)
21. Graves, A., Mohamed, A.R., Hinton, G.: Speech recognition with deep recurrent neural networks. In: 2013 IEEE International Conference on Acoustics, Speech and Signal Processing, pp. 6645–6649. IEEE, New York (2013)
22. Gulgec, N.S., Takáč, M., Pakzad, S.N.: Convolutional neural network approach for robust structural damage detection and localization. J. Comput. Civ. Eng. **33**(3), 04019005 (2019)
23. Li, L., Jamieson, K., DeSalvo, G., Rostamizadeh, A., Talwalkar, A.: Hyperband: a novel bandit-based approach to hyperparameter optimization. J. Mach. Learn. Res. **18**(1), 6765–6816 (2017)
24. Kingma, D. P., Ba, J.: Adam: a method for stochastic optimization. arXiv preprint arXiv:1412.6980 (2014)
25. Werbos, P.J.: Backpropagation through time: what it does and how to do it. Proc. IEEE. **78**(10), 1550–1560 (1990)

Operational Modal Analysis and Finite Element Model Updating of a 53-Story Building

Onur Avci, Khalid Alkhamis, Osama Abdeljaber, and Mohammed Hussein

Abstract This paper presents Operational Modal Analysis (OMA) and Finite Element (FE) model updating of a tall structure. Located in the West Bay area of Doha (Qatar), the structure was constructed between 2012 and 2016. It is a reinforced concrete building with shear wall cores located towards the center of the building plan. With 53 stories above the ground and 2 stories below ground, the 230 m (755 ft) tall building is being used for residential and hotel purposes. The material presented here is arguably the first published work on large-scale dynamic testing of a civil structure in Qatar. The wireless sensors used for testing are state-of-the-art equipment that can capture very low frequencies, something that cannot be accomplished with most of the conventional accelerometers available in the market.

Keywords Operational modal analysis · Finite element modeling · Structural dynamics · Tall structures · Model updating

1 Introduction

Operational Modal Analysis (OMA) and Finite Element (FE) model updating of a 53-story tower are presented in this paper. The structure was constructed between 2012 and 2016 and it is one of the tallest buildings in Qatar (Fig. 1a). The lateral load resisting system of the structure is concrete shear walls connected to reinforced concrete columns placed along the perimeter of the building with beams (Fig. 1b). The building has 53 stories above ground and 2 stories below ground, and it is being used for residential and hotel purposes. The building is 230 m (755 ft) tall. While this work is arguably the first published work on large-scale dynamic testing of a civil structure in Qatar, the wireless sensors used in this OMA work are state-of-the-art accelerometers sensitive enough to capture very low frequency modes of the structure. When the authors initially attempt to use standard wired sensors, it was realized that the conventional accelerometers which were successfully used in laboratory environment were not able to recognize the modal properties of lower frequency modes of the tower.

Basically, OMA is simply a modal testing method through which the dynamic characteristics of a structure is estimated based on the dynamic response under the ambient conditions [1–6]. This means the ambient forces and excitations the structure is subjected to during the operational use (wind, earthquake, traffic, machinery, human-excitations) are contributing to the analysis. The OMA methodology is often referred to as "ambient vibration modal identification" or "output-only modal analysis" by researchers [7–11]. Large structures like civil engineering infrastructure have been a good match for OMA use since it would be more difficult to excite a massive civil structure with artificial dynamic loading, than a much smaller laboratory structure. Therefore, the "output-only" component of OMA fits well for large structures. Researchers and engineers have been using the procedures developed for structural damage detection (SDD) and structural health monitoring (SHM) in OMA applications [12–17]. OMA results have always been found useful to verify the results of the computerized

O. Avci (✉)
Civil, Construction and Environmental Engineering, Iowa State University, Ames, IA, USA
e-mail: oavci@vt.edu; oavci@iastate.edu

K. Alkhamis · M. Hussein
Department of Civil and Architectural Engineering, Qatar University, Doha, Qatar
e-mail: mhussein@qu.edu.qa

O. Abdeljaber
Department of Building Technology, Linnaeus University, Växjö, Sweden
e-mail: osama.abdeljaber@lnu.se

K. Grimmelsman (ed.), *Dynamics of Civil Structures, Volume 2*, Conference Proceedings of the Society for Experimental Mechanics Series, https://doi.org/10.1007/978-3-030-77143-0_9

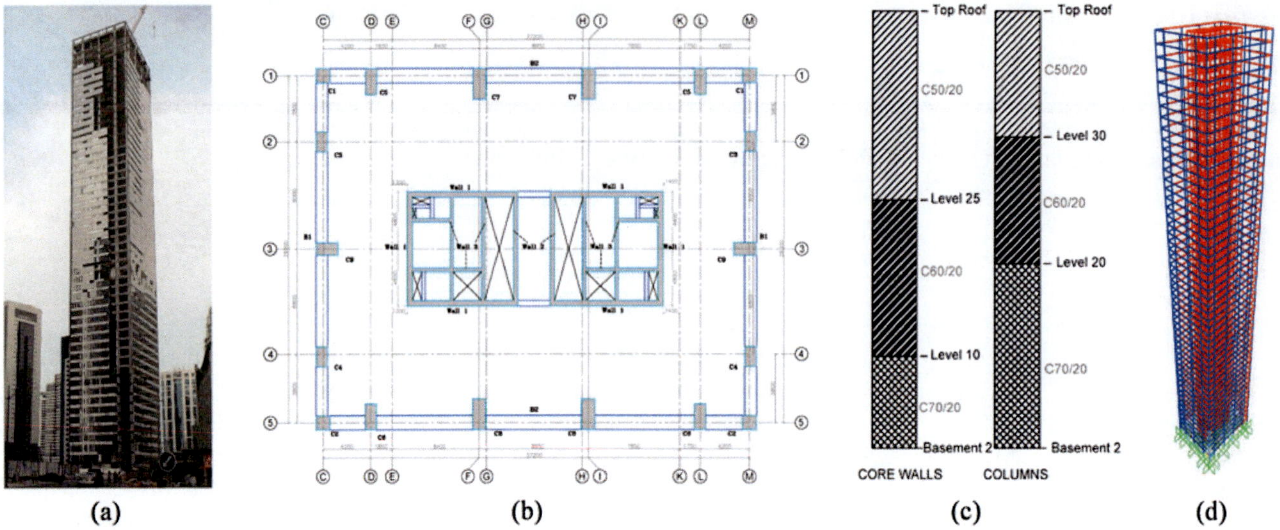

Fig. 1 (**a**) Tower structure; (**b**) typical plan view; (**c**) variation of concrete strength along the height; (**d**) 3D FE model

simulations since the modal characteristics obtained from FE models are methodological procedures full of approximations based on the geometry of structural members, assumed material properties, and eigenvector analysis. Yet, in civil engineering applications, it is not easy to create a perfect FE model that can successfully predict all aspects of static and dynamic characteristics of a structure [18–21] since the overall structural behavior is a function of combination many parameters, assumptions, and interactions (e.g., member-to-member connections, structural mass, amplitude-dependent damping, creep-shrinkage effects, stiffness of cladding and/or façade elements, non-structural components, soil-structure interaction) [22–25]. The OMA results are not only compared to the Finite Element (FE) model results, but they are also being used in updating the FE models of structures. On another note, it is known that dynamic excitations applied at or around the modal frequencies have increasing effect of dynamic loads applied on the structures which are causing discomfort for the tenants and structural damage to the buildings. Therefore, monitoring and keeping track of variations on the dynamic properties of structures through OMA is valuable and pertinent for life-cycle assessment and structural integrity of civil infrastructure [26–31]. OMA results have also been reported to be utilized at various stages of construction to monitor and verify the performance, requirements, and assumptions at the site.

All elevated floors of the tested building are 300 mm thick flat slabs with drop panels. Floor-to-floor height is typically 3.5 m for all levels except the lobby and mechanical floors. The lateral load resisting elements (columns and shear walls) are connected to each other with beams varying in size (1.20 m × 0.45 m) and (1.00 m × 0.45 m) to create a diaphragm effect. During the OMA procedures, the building was still under construction with 90% of the work had been completed. Different grades and strength of concrete were used throughout the height of the tower for both core walls and columns (Fig. 1c), which is reflected in the FE model (Fig. 1d). Eigenvector analysis was chosen in the commercial FE software package, and first ten modes are targeted to be studied and then compared to the OMA procedures. The first ten modes predicted by the FE model are shown in Fig. 2.

2 Operational Modal Analysis

For the 53-story high structure, to capture the modes below 0.5–0.6 Hz range, ten TROMINO wireless sensors were used for data collection. In wireless sensor networks, time synchronization demands specific solutions whereas it is a standard task for wired sensors. Each node in wireless sensor networks has its own analog-to-digital (ADC) converter. The performance of the sensors used in this study had already been verified on various other OMA projects on large civil infrastructure. The sensor uses three channels each assigned for three orthogonal directions. Data was collected for a period of 90 min at a rate of 128 samples per second. First and last 10 min were removed from the 90 min to focus on the data collected in the middle chunk of the recordings. Two sensors were placed on the opposite corners of the instrumented floors. The instrumented floors were the 1st, 13th, 25th, 37th, and the 48th levels (a total of five levels on the tower) as shown in Fig. 3a. The first reference sensor was placed on the first floor. All collected data was uploaded to MEScope software to extract modal features. A sample data

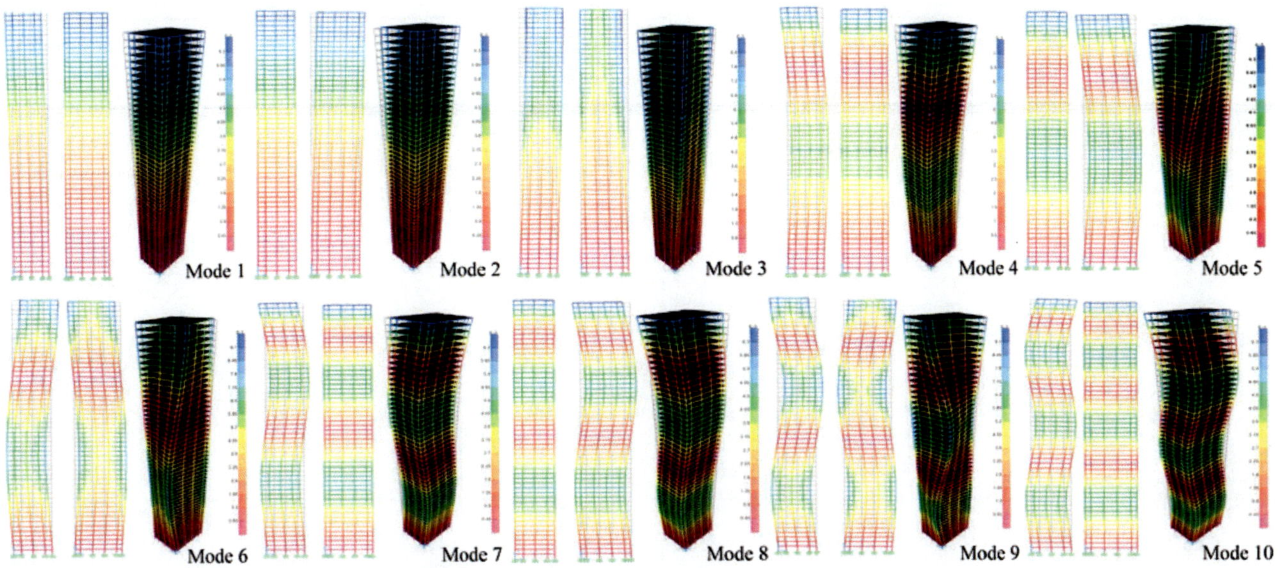

Fig. 2 First ten mode shapes predicted by the FE model

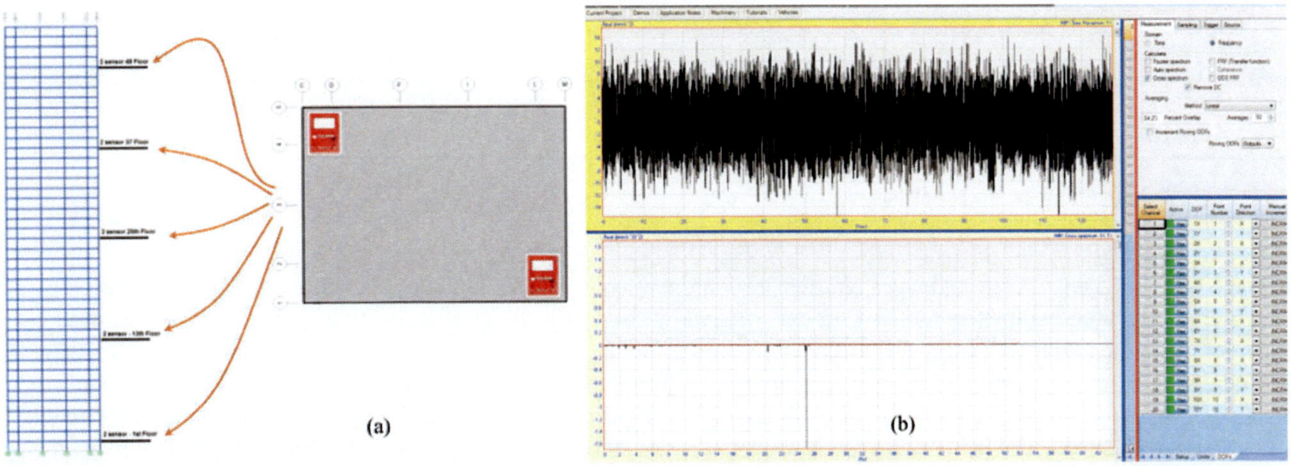

Fig. 3 (**a**) Sensor placement at five levels; (**b**) sample data acquisition window in MEScope

acquisition window is shown in Fig. 3b. Sample power spectral density windows captured in MEScope are shown in Fig. 4. The curve fitted mode shapes, mode descriptions, modal damping ratios, and modal frequencies per MEScope analysis are presented in Fig. 5.

3 Finite Element Model Updating

After the OMA procedures, natural frequencies obtained from OMA are compared to the FE natural frequency predictions. The error between the two are calculated and presented in Table 1. It is noted that the average absolute error is 15.3%, and when the torsional modes (modes 3, 6, and 9) are excluded, the average absolute error is 15.5% for the remaining modes. The Modal Assurance Criterion (MAC) plot is also presented in Table 1. The MAC is commonly used as a statistical indicator by researchers and practitioners. It is used simply to determine the similarity of two mode shapes. It is reported to be sensitive to relatively larger variations, and partially insensitive to relatively smaller variations in mode shapes [32–37]. Based on the MAC plot in Table 1, it is observed that the correlation for modes 3, 7, 8, and 10 needs improvement. Therefore, it is decided that the FE model is updated in an attempt to improve the correlation between the OMA and FE model natural frequencies. Updating FE models has been discussed extensively in the literature, and modifications on various parameters such as mass,

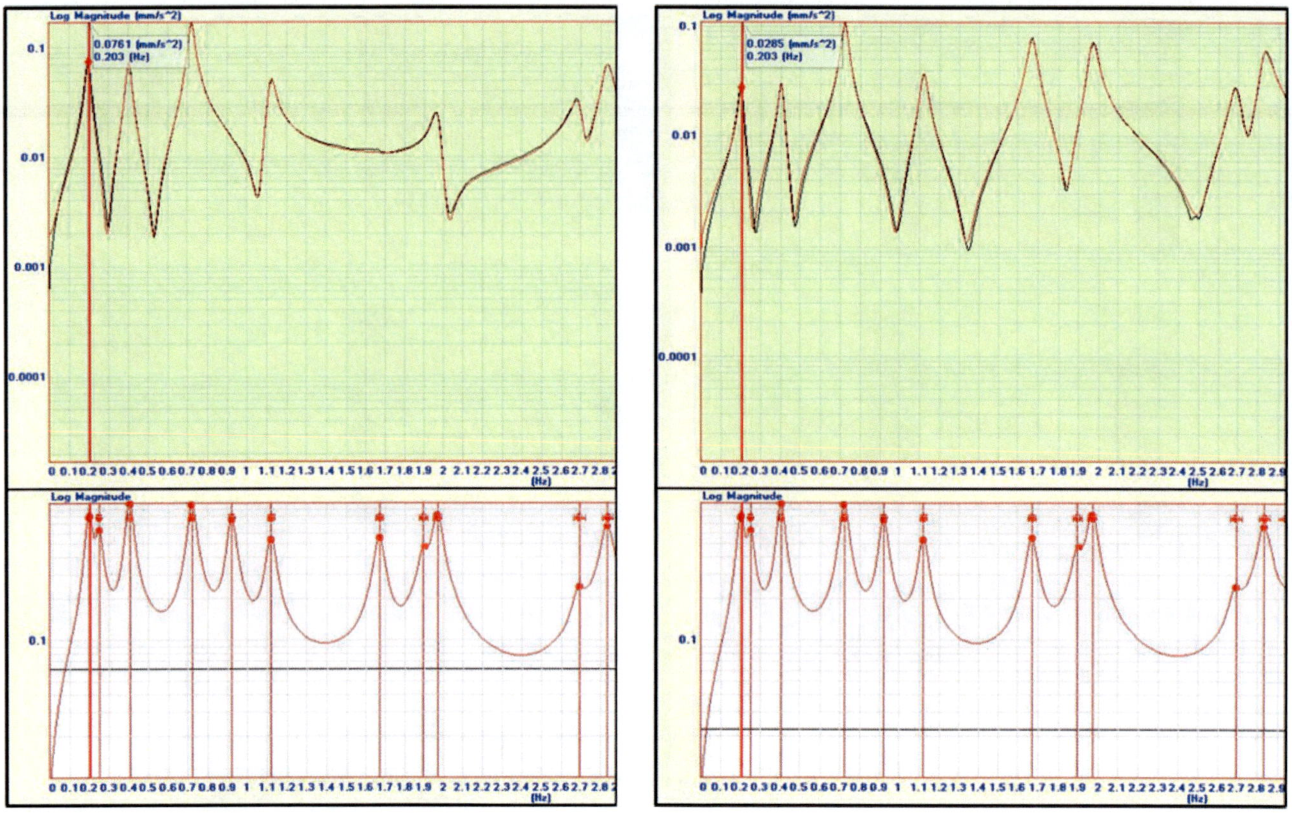

Fig. 4 Power Spectral Density windows in MEScope

elastic modulus, moment of inertia, fixities, cracked sections, and loads on the FE models are accepted ways of approaching the measured structural parameters [38–46]. In experimental testing, regardless of static and dynamic properties, the FE models are calibrated with the measured results collected on the actual structures [47–55]. The first attempt of updating the FE model here in this work was decreasing the self-weight of the building. A drastic decrease of 42% is applied on the concrete mass to observe the outcome, and the resulting (updated) FEM-predicted frequencies and the corresponding MAC plot are presented in Table 2. With this FE model, it is noted that the average absolute error is 10.4%, and when the torsional modes (modes 3, 6, and 9) are excluded, the average absolute error is 5.9% for the remaining modes. While there are enhancements on the MAC plot on Table 2 when compared to Table 1, it is decided that the FE model is updated again, to see the effect of changes on the modulus of elasticity. Therefore, this time the elastic modulus is increased drastically (70%) to observe the outcome. The resulting (updated) FEM-predicted frequencies and the corresponding MAC plot are shown in Table 3. Again, there are improvements observed on the MAC plot on Table 3. With this FE model, it is noted that the average absolute error is 9.5%, and when the torsional modes (modes 3, 6, and 9) are excluded, the average absolute error is 5.8% for the remaining modes. The FE model is then updated with modifications on both mass and elastic modulus, simultaneously. This time the update is done by relatively reasonable numbers: 22% decrease in mass and 25% increase in elastic modulus. In addition to these modifications in the mass and elastic modulus, for this FE model update, the torsional fixities of the columns and walls are decreased by 50%. The updated results are presented in Table 4. Based on these changes on the FE model, it is observed that the MAC plot is further improved; the average absolute error has come down to 7.7%, and when the torsional modes (modes 3, 6, and 9) are excluded, the average absolute error has come down all the way to 5.1% for the remaining modes.

4 Conclusions

In this paper, Operational Modal Analysis (OMA) and Finite Element (FE) model updating of a 53-story structure in Qatar is presented. The 230 m tall structure was tested with wireless sensors. Focusing on the first ten modes of the structure, OMA procedures were completed. Meanwhile, natural frequencies of the structure were predicted with a commercial finite

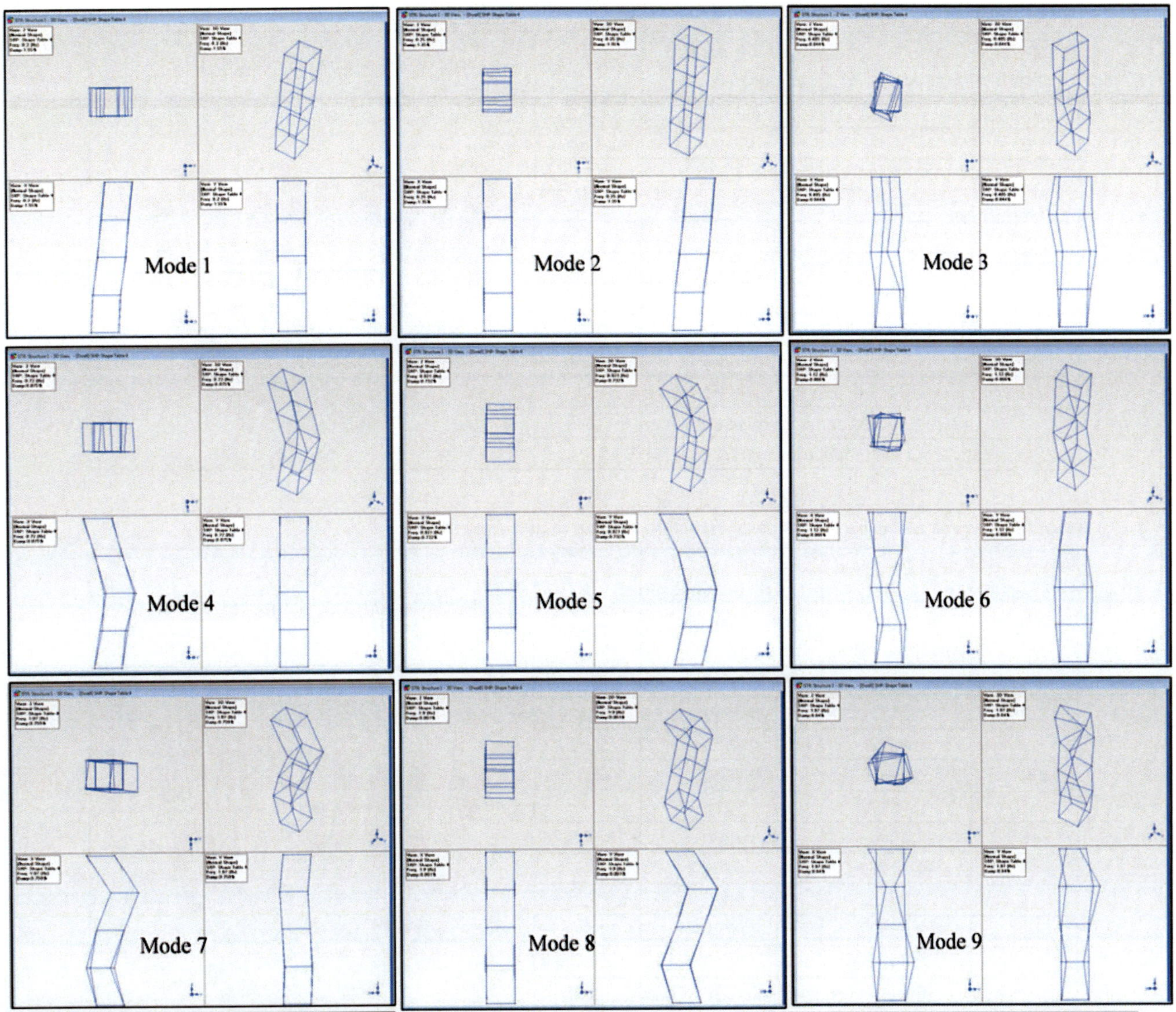

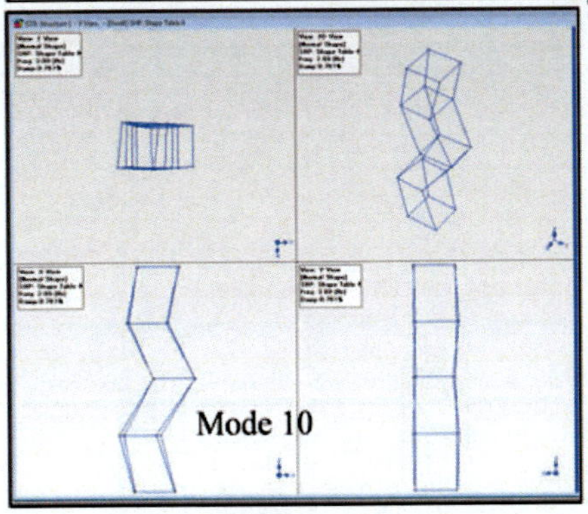

Mode number	Mode description	OMA measured frequency (Hz)	OMA measured damping ratio (%)
1	Translation in E-W direction	0.200	1.550
2	Translation in N-S direction	0.250	1.350
3	Torsional	0.405	0.844
4	Translation in E-W direction	0.720	0.875
5	Translation in N-S direction	0.921	0.732
6	Torsional	1.120	0.866
7	Translation in E-W direction	1.670	0.759
8	Translation in N-S direction	1.900	0.801
9	Torsional	1.970	0.640
10	Translation in E-W direction	2.690	0.761

Fig. 5 MEScope results for mode descriptions, mode shapes, modal frequencies, modal damping ratios

element modeling software. Since the resulting Modal Assurance Criterion (MAC) plot needed improvement, the FE model was updated/calibrated several times by decreasing the mass, increasing the elastic modulus, and partially releasing the torsional restraint on columns and walls. With each FE model update, it is observed that the MAC plots were improved, and the errors between the measured and predicted frequencies decreased. As such, a successful FE model updating procedure has been demonstrated. It is also important to note that the work presented here is arguably the first published work on large-scale dynamic testing of a civil infrastructure in Qatar.

Table 1 Measured and FEM-predicted natural frequencies

Mode number	FEM Predicted Frequency (Hz)	OMA Measured Frequency (Hz)	Error (%)	Modal Assurance Criterion (MAC) Plot
1	0.1504	0.2000	-24.8	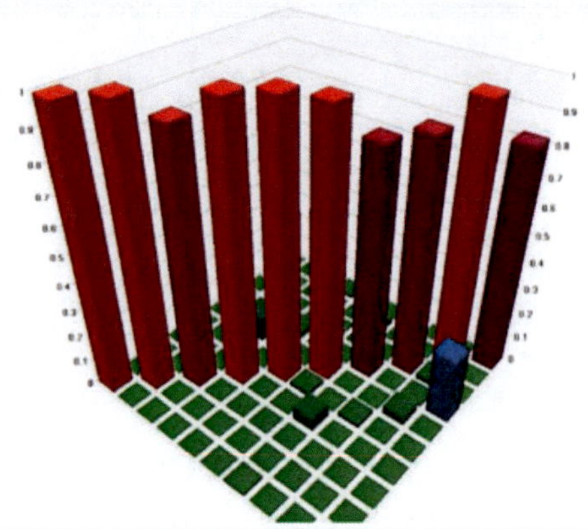
2	0.2062	0.2500	-17.5	
3	0.4724	0.4050	16.6	
4	0.6059	0.7200	-15.8	
5	0.7501	0.9210	-18.6	
6	1.2939	1.1200	15.5	
7	1.4357	1.6700	-14.0	
8	1.6165	1.9000	-14.9	
9	2.2104	1.9700	12.2	
10	2.6090	2.6900	-3.0	
Average absolute error (%)			15.3	
Average abs. error excluding torsional modes (%)			15.5	

Table 2 Measured and updated FEM-predicted natural frequencies (decrease in mass)

Mode number	FEM Predicted Frequency per model update (modified mass)	OMA Measured Frequency (Hz)	Error (%)	Modal Assurance Criterion (MAC) Plot
1	0.2009	0.2000	0.4	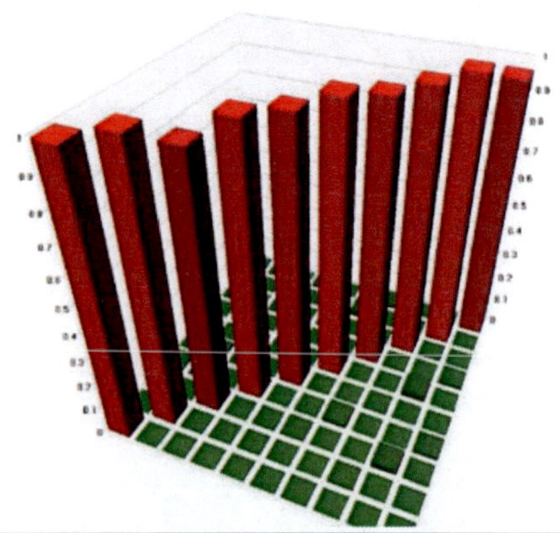
2	0.2555	0.2500	2.2	
3	0.5232	0.4050	29.2	
4	0.7535	0.7200	4.7	
5	0.9877	0.9210	7.2	
6	1.3288	1.1200	18.6	
7	1.8112	1.6700	8.5	
8	2.0745	1.9000	9.2	
9	2.2532	1.9700	14.4	
10	2.9422	2.6900	9.4	
Average absolute error (%)			10.4	
Average abs. error excluding torsional modes (%)			5.9	

Table 3 Measured and updated FEM-predicted natural frequencies (decrease in mass)

Mode number	FEM Predicted Frequency per model update (Modified elastic modulus)	OMA Measured Frequency (Hz)	Error (%)	Modal Assurance Criterion (MAC) Plot
1	0.1961	0.2000	-2.0	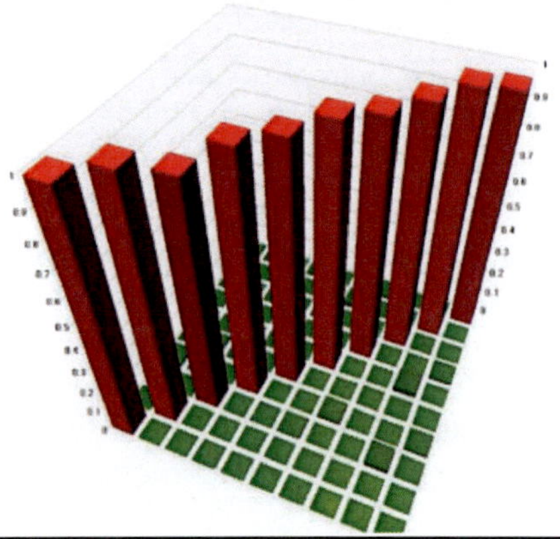
2	0.2588	0.2500	3.5	
3	0.5010	0.4050	23.7	
4	0.7550	0.7200	4.9	
5	0.9780	0.9210	6.2	
6	1.3170	1.1200	17.6	
7	1.7920	1.6700	7.3	
8	2.0866	1.9000	9.8	
9	2.2321	1.9700	13.3	
10	2.8785	2.6900	7.0	
Average absolute error (%)			9.5	
Average abs. error excluding torsional modes (%)			5.8	

Table 4 Measured and updated FEM-predicted natural frequencies (decrease in mass + increase in elastic modulus + partial torsional release on columns and walls)

Mode number	FEM Predicted Frequency per model update (Modified mass and elastic modulus)	OMA Measured Frequency (Hz)	Error (%)	Modal Assurance Criterion (MAC) Plot
1	0.1865	0.2000	-6.8	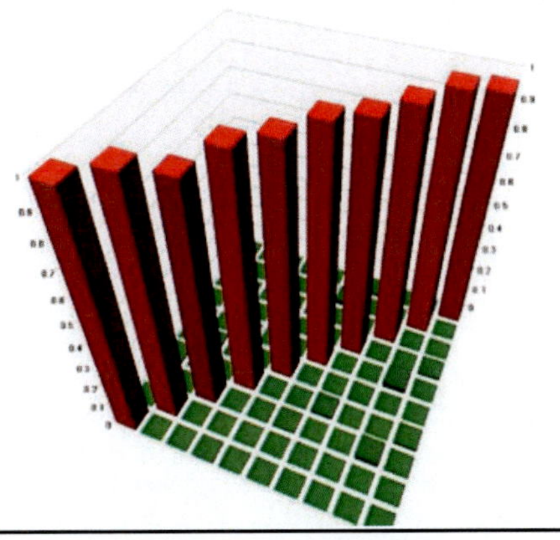
2	0.2434	0.2500	-2.6	
3	0.4713	0.4050	16.4	
4	0.7350	0.7200	2.1	
5	0.9730	0.9210	5.6	
6	1.2758	1.1200	13.9	
7	1.7559	1.6700	5.1	
8	2.0142	1.9000	6.0	
9	2.1841	1.9700	10.9	
10	2.8891	2.6900	7.4	
Average absolute error (%)			7.7	
Average abs. error excluding torsional modes (%)			5.1	

References

1. Reynders, E.: System identification methods for (operational) modal analysis: review and comparison. Arch. Comput. Methods Eng. (2012). https://doi.org/10.1007/s11831-012-9069-x
2. Kim, D., Oh, B.K., Park, H.S., Shim, H.B., Kim, J.: Modal identification for high-rise building structures using orthogonality of filtered response vectors. Comput. Civ. Infrastruct. Eng. (2017). https://doi.org/10.1111/mice.12310
3. Cunha, A., Caetano, E., Magalhães, F.: Operational modal analysis for testing and continuous monitoring of bridges and special structures. In: Concr. Repair, Rehabil. Retrofit. III - Proc. 3rd Int. Conf. Concr. Repair, Rehabil. Retrofit. ICCRRR 2012, p. 2012
4. Gentile, C., Saisi, A.: Operational modal testing of historic structures at different levels of excitation. Constr. Build. Mater. (2013). https://doi.org/10.1016/j.conbuildmat.2013.01.013
5. Ghahari, S.F., Abazarsa, F., Ghannad, M.A., Taciroglu, E.: Response-only modal identification of structures using strong motion data. Earthq. Eng. Struct. Dyn. (2013). https://doi.org/10.1002/eqe.2268
6. Benedettini, F., Gentile, C.: Operational modal testing and FE model tuning of a cable-stayed bridge. Eng. Struct. (2011). https://doi.org/10.1016/j.engstruct.2011.02.046
7. Brownjohn, J.M.W., Magalhaes, F., Caetano, E., Cunha, A.: Ambient vibration re-testing and operational modal analysis of the Humber Bridge. Eng. Struct. (2010). https://doi.org/10.1016/j.engstruct.2010.02.034
8. Gentile, C., Saisi, A., Cabboi, A.: Structural identification of a masonry tower based on operational modal analysis. Int. J. Archit. Herit. (2015). https://doi.org/10.1080/15583058.2014.951792
9. Aenlle, M.L., Brincker, R.: Modal scaling in operational modal analysis using a finite element model. Int. J. Mech. Sci. (2013). https://doi.org/10.1016/j.ijmecsci.2013.09.003
10. Masjedian, M.H., Keshmiri, M.: A review on operational modal analysis researches: classification of methods and applications. In: IOMAC 2009 - 3rd Int. Oper. Modal Anal. Conf. (2009)
11. Omrani, R., Hudson, R.E., Taciroglu, E.: Parametric identification of nondegrading hysteresis in a laterally and torsionally coupled building using an unscented Kalman filter. J. Eng. Mech. (2013). https://doi.org/10.1061/(asce)em.1943-7889.0000498
12. Brownjohn, J.M., Pan, T.-C.: Identifying loading and response mechanisms from ten years of performance monitoring of a tall building. J. Perform. Constr. Facil. (2008). https://doi.org/10.1061/(asce)0887-3828(2008)22:1(24)
13. Mohanty, P., Rixen, D.J.: Operational modal analysis in the presence of harmonic excitation. J. Sound Vib. (2004). https://doi.org/10.1016/S0022-460X(03)00485-1
14. Devriendt, C., De Sitter, G., Vanlanduit, S., Guillaume, P.: Operational modal analysis in the presence of harmonic excitations by the use of transmissibility measurements. Mech. Syst. Signal Process. (2009). https://doi.org/10.1016/j.ymssp.2008.07.009
15. Ni, Y., Lu, X., Lu, W.: Operational modal analysis of a high-rise multi-function building with dampers by a Bayesian approach. Mech. Syst. Signal Process. (2017). https://doi.org/10.1016/j.ymssp.2016.10.009
16. Ramos, L.F., Marques, L., Lourenço, P.B., De Roeck, G., Campos-Costa, A., Roque, J.: Monitoring historical masonry structures with operational modal analysis: two case studies. Mech. Syst. Signal Process. (2010). https://doi.org/10.1016/j.ymssp.2010.01.011
17. Reynders, E., Houbrechts, J., De Roeck, G.: Fully automated (operational) modal analysis. Mech. Syst. Signal Process. (2012). https://doi.org/10.1016/j.ymssp.2012.01.007

18. Avci, O., Al-Smadi, Y.M.: Unreinforced masonry façade assessment of a historic building for excessive displacements due to a nearby subway construction. Pract. Period. Struct. Des. Constr. (2019). https://doi.org/10.1061/(asce)sc.1943-5576.0000399
19. Avci, O., Al Nouss, M.: Seismic assessment of existing lowrise and midrise reinforced concrete buildings using the 2014 Qatar construction specification. J. Archit. Eng. (2018). https://doi.org/10.1061/(asce)ae.1943-5568.0000331
20. Avci, O.: Web-Crippling Strength of Multi-web Cold-Formed Steel Deck Sections Subjected to End One Flange (EOF) Loading. Virginia Polytechnic Institute and State University, Virginia (2002)
21. Avci, O.: Nonlinear damping in floor vibrations serviceability: verification on a laboratory structure. Conf. Proc. Soc. Exp. Mech. Ser. (2017). https://doi.org/10.1007/978-3-319-54777-0_18
22. Avci, O.: Amplitude-dependent damping in vibration serviceability: case of a laboratory footbridge. J. Archit. Eng. (2016). https://doi.org/10.1061/(asce)ae.1943-5568.0000211
23. Avci, O.: Modal parameter variations due to joist bottom chord extension installations on laboratory footbridges. J. Perform. Constr. Facil. (2015). https://doi.org/10.1061/(asce)cf.1943-5509.0000635
24. Avci, O.: Retrofitting Steel Joist Supported Footbridges for Improved Vibration Response Struct. Congr. 2012 - Proc. 2012 (2012). https://doi.org/10.1061/9780784412367.041
25. Al-Smadi, Y.M., Bhargava, A., Avci, O., Elmorsi, M.: Design of experiments study to obtain a robust 3D computational bridge model. Conf. Proc. Soc. Exp. Mech. Ser. (2012). https://doi.org/10.1007/978-1-4614-2413-0_29
26. Rainieri, C., Fabbrocino, G.: Operational Modal Analysis of Civil Engineering Structures. Springer, Cham (2014). https://doi.org/10.1007/978-1-4939-0767-0
27. Zhang, L., Brincker, R., Andersen, P.: An overview of operational modal analysis: major development and issues. In: Proc. 1st Int. Oper. Modal Anal. Conf. IOMAC 2005 (2005)
28. Batel, M.: Operational modal analysis - another way of doing modal testing. Sound Vib. (2002)
29. Zhang, F.L., Ni, Y.Q., Ni, Y.C., Wang, Y.W.: Operational modal analysis of Canton Tower by a fast frequency domain Bayesian method. Smart Struct. Syst. (2016). https://doi.org/10.12989/sss.2016.17.2.209
30. Zhang, F.L., Yang, Y.P., Xiong, H.B., Yang, J.H., Yu, Z.: Structural health monitoring of a 250-m super-tall building and operational modal analysis using the fast Bayesian FFT method. Struct. Control Health Monit. (2019). https://doi.org/10.1002/stc.2383
31. Karbhari, V.M., Guan, H., Sikorsky, C.: Operational modal analysis for vibration-based structural health monitoring of civil structures. Struct. Health Monit. Civ. Infrastruct. Syst. (2009). https://doi.org/10.1533/9781845696825.1.213
32. Fotsch, D., Ewins, D.J.: Application of MAC in the frequency domain. In: Proc. Int. Modal Anal. Conf. - IMAC (2000)
33. Vacher, P., Jacquier, B., Bucharles, A.: Extensions of the MAC criterion to complex modes. In: Proc. ISMA 2010 - Int. Conf. Noise Vib. Eng. Incl. USD 2010 (2010)
34. Rigner, L.: Modal assurance criteria value for two orthogonal modal vectors. In: Proc. Int. Modal Anal. Conf. - IMAC (1998)
35. Pastor, M., Binda, M., Harčarik, T.: Modal assurance criterion. Procedia Eng. (2012). https://doi.org/10.1016/j.proeng.2012.09.551
36. Allemang, R.J.: The modal assurance criterion - twenty years of use and abuse. Sound Vib. (2003)
37. Lei, S., Mao, K., Li, L., Xiao, W., Li, B.: Sensitivity analysis of modal assurance criteria of damped systems. J. Vib. Control. (2017). https://doi.org/10.1177/1077546315582122
38. Ghahari, S.F., Abazarsa, F., Avci, O., Çelebi, M., Taciroglu, E.: Blind identification of the Millikan Library from earthquake data considering soil-structure interaction. Struct. Control Health Monit. (2016). https://doi.org/10.1002/stc.1803
39. Davis, B., Avci, O.: Simplified Vibration Response Prediction for Slender Monumental Stairs Struct. Congr. 2014 - Proc. (2014). https://doi.org/10.1061/9780784413357.223
40. Davis, B., Avci, O.: Simplified vibration serviceability evaluation of slender monumental stairs. J. Struct. Eng. (2015). https://doi.org/10.1061/(asce)st.1943-541x.0001256
41. Bhargava, A., Isenberg, J., Feenstra, P.H., Al-Smadi, Y., Avci, O.: Vibrations assessment of a hospital floor for a magnetic resonance imaging unit (MRI) replacement. In: Struct. Congr. 2013 Bridg. Your Passion with Your Prof. - Proc. 2013 Struct. Congr. (2013). https://doi.org/10.1061/9780784412848.212
42. Barsottelli, M., Avci, O.: Fundamentals of highway bridge demolition. In: Struct. Congr. 2013 Bridg. Your Passion with Your Prof. - Proc. 2013 Struct. Congr. (2013). https://doi.org/10.1061/9780784412848.060
43. Avci, O., Setareh, M., Murray, T.M.: Vibration testing of joist supported footbridges. In: Struct. Congr. 2010 (2010). https://doi.org/10.1061/41130(369)80
44. Avci, O., Murray, T.M.: Effect of bottom chord extensions on the static flexural stiffness of open-web steel joists. J. Perform. Constr. Facil. (2012). https://doi.org/10.1061/(asce)cf.1943-5509.0000262
45. Avci, O., Mattingly, J., Luttrell, L.D., Samuel Easterling, W.: Roof diaphragm strength and stiffness. In: Seventeenth Int. Spec. Conf. Cold-Formed Steel Struct. Recent Res. Dev. Cold-Formed Steel Des. Constr. (2005)
46. O. Avci, L.D. Luttrell, J. Mattingly, W.S. Easterling, Diaphragm shear strength and stiffness of aluminum roof panel assemblies, Thin-Walled Struct. (2016). doi:https://doi.org/10.1016/j.tws.2016.04.019
47. Avci, O., Easterling, W.S.: WEB crippling strength of multi-web steel deck sections subjected to end one flange loading. In: Int. Spec. Conf. Cold-Formed Steel Struct. Recent Res. Dev. Cold-Formed Steel Des. Constr. (2002)
48. Avci, O., Easterling, W.S.: Web crippling strength of steel deck subjected to end one flange loading. J. Struct. Eng. (2004). https://doi.org/10.1061/(asce)0733-9445(2004)130:5(697)
49. Avci, O., Davis, B.: A Study on Effective Mass of One Way Joist Supported Systems Struct. Congr. 2015 - Proc. 2015 (2015). https://doi.org/10.1061/9780784479117.073
50. Avci, O., Bhargava, A., Nikitas, N., Inman, D.J.: Vibration annoyance assessment of train induced excitations from tunnels embedded in rock. Sci. Total Environ. (2020). https://doi.org/10.1016/j.scitotenv.2019.134528
51. Avci, O., Bhargava, A., Nikitas, N., Inman, D.J.: Vibrations assessment of existing building foundations due to moving trains in underground tunnels. Conf. Proc. Soc. Exp. Mech. Ser. (2021). https://doi.org/10.1007/978-3-030-47634-2_8
52. Avci, O., Bhargava, A., Al-Smadi, Y., Isenberg, J.: Vibrations serviceability of a medical facility floor for sensitive equipment replacement: evaluation with sparse in situ data. Pract. Period. Struct. Des. Constr. (2019). https://doi.org/10.1061/(asce)sc.1943-5576.0000404

9. Ramsay, J.O., Silverman, B.W.: Applied Functional Data Analysis: Methods and Case Studies. Springer, New York (2007)
10. Zhou, R.R., Serban, N., Gebraeel, N.: Degradation modeling applied to residual lifetime prediction using functional data analysis. Ann. Appl. Stat. **5**(2B), 1586–1610 (2011). https://doi.org/10.1214/10-AOAS448
11. Sun, J., Liao, H., Upadhyaya, B.R.: A robust functional-data-analysis method for data recovery in multichannel sensor systems. IEEE Trans. Cybern. **44**(8), 1420–1431 (2014). https://doi.org/10.1109/TCYB.2013.2285876
12. Li, T.: From zero crossings to quantile-frequency analysis of time series with an application to nondestructive evaluation. (2019). https://doi.org/10.1002/asmb.2499
13. Chen, Z., Li, H., Bao, Y.: Analyzing and modeling inter-sensor relationships for strain monitoring data and missing data imputation: a copula and functional data-analytic approach. Struct. Health Monit. **18**(4), 1168–1188 (2019). https://doi.org/10.1177/1475921718788703
14. Chen, Z., Bao, Y., Li, H., Spencer, B.F.: A novel distribution regression approach for data loss compensation in structural health monitoring. Struct. Health Monit. **17**(6), 1473–1490 (2018). https://doi.org/10.1177/1475921717745719
15. Chen, Z., Bao, Y., Li, H., Spencer, B.F.: LQD-RKHS-based distribution-to-distribution regression methodology for restoring the probability distributions of missing SHM data. Mech. Syst. Signal Process. **121**, 655–674 (2019). https://doi.org/10.1016/j.ymssp.2018.11.052
16. Chen, Z., Bao, Y., Tang, Z., Chen, J., Li, H.: Clarifying and quantifying the geometric correlation for probability distributions of inter-sensor monitoring data: a functional data analytic methodology. Mech. Syst. Signal Process. **138**, 106540 (2020). https://doi.org/10.1016/j.ymssp.2019.106540
17. Merve, U.: SERIES RC Walls Database. https://datacenterhub.org/resources/355 (2015)
18. Momeni, H., Dolatshahi, K.M.: Predictive equations for drift ratio and damage assessment of RC shear walls using surface crack patterns. Eng. Struct. **190**, 410–421 (2019). https://doi.org/10.1016/j.engstruct.2019.04.018
19. Momeni, H.: Probabilistic Damage Assessment of Reinforced Concrete Shear Walls Using Crack Pattern. Sharif University of Technology, Tehran (2019)
20. Lu, X., Zhou, Y., Yang, J., Qian, J., Song, C., Wang, Y.: Shear Wall Database. NEEShub Database (2010)
21. Tran, T.A., Wallace, J.W.: Experimental study of nonlinear flexural and shear deformations of reinforced concrete structural walls. In: Proceedings, 15th World Conference on Earthquake Engineering, Lisbon, Portugal (2012)
22. Hosseini, S.A., Kheyroddin, A., Mastali, M.: An experimental investigation into the impacts of eccentric openings on the in-plane behavior of squat RC shear walls. Eng. Struct. **197**, 109410 (2019)
23. Ebrahimkhanlou, A., Choi, J., Hrynyk, T.D., Salamone, S., Bayrak, O.: Detection of the onset of delamination in a post-tensioned curved concrete structure using hidden Markov modeling of acoustic emissions. In: Sensors and Smart Structures Technologies for Civil, Mechanical, and Aerospace Systems 2018, vol. 10598, p. 1059821 (2018)
24. Ebrahimkhanlou, A., Choi, J., Hrynyk, T.D., Salamone, S., Bayrak, O.: Acoustic emission monitoring of containment structures during post-tensioning. Eng. Struct. **209**, 109930 (2020)
25. Ebrahimkhanlou, A., et al.: Acoustic emission monitoring of strengthened steel bridges: inferring the mechanical behavior of post-installed shear connectors. In: Nondestructive Characterization and Monitoring of Advanced Materials, Aerospace, Civil Infrastructure, and Transportation XIII, vol. 10971, p. 109710H (2019)
26. Escarcega, M., Bradley, S., Randall, P., Campos, G., Strebe, L., Momeni, H., Ebrahimkhanlou, A.: Corrosion monitoring and mitigation in concrete structures using a 3D printing and internet of things approach. In: Sensors and Smart Structures Technologies for Civil, Mechanical, and Aerospace Systems, vol. 11591, p. 115910X (2021). https://doi.org/10.1117/12.2583360
27. Momeni, H., Basereh, S., Okumus, P., Ebrahimkhanlou, A.: Surface crack detection in concrete structures using video processing techniques. Health Monit. Struct. Bio. Syst. XV. **11593**, 115932B (2021). https://doi.org/10.1117/12.2583323

An Overview of Deep Learning Methods Used in Vibration-Based Damage Detection in Civil Engineering

Onur Avci, Osama Abdeljaber, and Serkan Kiranyaz

Abstract This paper presents a brief overview of vibration-based damage identification studies based on Deep Learning (DL) in civil engineering structures. The presence, type, size, and propagation of structural damage on civil infrastructure have always been a topic of research. In the last couple of decades, there has been a significant shift in the damage detection paradigm when the advancements in sensing and computing technologies met with the ever-expanding use of artificial neural network algorithms. Machine-Learning (ML) tools enabled researchers to implement more feasible and faster tools in damage detection applications. When an artificial neural network has more than three layers, it is typically considered as a "deep" learning network. Being an important accomplishment of the ML era, DL tools enable complex systems which are made of several layers to learn implementations of data with outstanding categorization and compartmentalization capability. In fact, with proper training, a DL tool can operate directly with the unprocessed raw data and help the algorithm produce output data. Competitive capabilities like this led DL algorithms perform very well in complicated problems by dividing a relatively large problem into much smaller and more manageable portions. Specifically for damage identification and localization on civil infrastructure, Convolutional Neural Networks (CNNs) and Unsupervised Pretrained Networks (UPNs) are the known DL tools published in the literature. This paper presents an overview of these studies.

Keywords Civil engineering structures · Damage identification · Damage localization · Infrastructure health · Deep learning

1 Introduction

Inspecting the overall performance of engineering structures has always been important for maintaining structural health. While the traditional way of damage monitoring on civil infrastructure has been through visual inspections [1–3], along with the advancements in sensors and monitoring technology, dynamic response of structures has started to be exploited for condition assessment [4–10] and serviceability evaluation [11–17] of structures. Along with the implementation of Machine Learning (ML) based procedures in structural damage detection (both nonparametric ML and parametric ML methods), it has been reported that both supervised ML procedures and unsupervised ML procedures need the step of feature extraction to be completed first, so that the input data is represented with reference to a certain number of manually selected features [18–20]. Even though some of the manually selected features operate very well for some specific cases, they might not necessarily work on other cases. Therefore, with the intention to keep away from manually selected features in complicated ML-based methods, Deep Learning (DL) methods were created. DL or "Deep Neural Network" is also referred to as "Deep Neural Learning" or "Representation Learning." It is a subset, indeed a special type of ML-based procedures. A great characteristic of the DL is, without programmer intervention, it can extract optimum input representation directly from the raw signal

O. Avci (✉)
Civil, Construction and Environmental Engineering, Iowa State University, Ames, IA, USA
e-mail: oavci@vt.edu; oavci@iastate.edu

O. Abdeljaber
Department of Building Technology, Linnaeus University, Växjö, Sweden
e-mail: osama.abdeljaber@lnu.se

S. Kiranyaz
Department of Electrical Engineering, Qatar University, Doha, Qatar
e-mail: mkiranyaz@qu.edu.qa

K. Grimmelsman (ed.), *Dynamics of Civil Structures, Volume 2*, Conference Proceedings of the Society for Experimental Mechanics Series, https://doi.org/10.1007/978-3-030-77143-0_10

while enhancing the classification accuracy. DL has the capability to learn by correlating the features per training, and then process the feature extraction step based on the training. The competitive attribute of the DL is the fact that it can learn in an unsupervised manner from unstructured data which enables it to tackle large problems by dividing it into smaller, more manageable problems. DL has been in use by numerous other fields but their applications in vibration-based damage detection of civil infrastructure only are discussed in this paper.

Standard Artificial Neural Networks (ANNs) typically comprise an input layer. This layer is followed by hidden layers, which are followed by an output layer. ANNs comprising multiple layers excluding input and output layers are accepted as "deep" networks. It can also be considered as when the number of hidden layers increase, then the network "depth" gets increased, to make the network a "Deep" one. For DL development, the study by Hinton and Salakhutdinov [21] is considered as a pioneer study. DL structures are divided into four parts in [22]: Recursive Neural Networks; Recurrent Neural Networks; Convolutional Neural Networks (CNNs); and Unsupervised Pretrained Networks (UPNs). Unsupervised Pretrained Networks are further categorized into Generative Adversarial Networks (GANs); Deep Belief Networks (DBNs); and Autoencoders (Deep Autoencoders). For damage detection of civil infrastructure, the DL tools utilized in the literature have been CNNs and UPNs. Therefore, this review includes these studies.

2 Use of UPNs in Damage Detection of Civil Infrastructure

Within the UPN categorization structure, Autoencoders are found to be the only network found in literature that was deployed for damage identification and localization of civil structures. Within DL use, the Deep Autoencoders comprise multiple hidden layers and original data description is more efficient per the learned features which enhances the classification performance. An ensemble classification technique was used in [23] and then in [24] utilizing weight majority voting. In these studies, the Autoencoder methodologies were referred to as Deep Neural Networks. Since the damage indicators are often sensitive to environmental conditions such as temperature changes, in [23], the proposed damage identification method addresses these factors by employing Couple Sparse Coding (CSC) and Autoencoders. Using analytical and experimental data from bridges, the methodology was verified with noisy data conditions, temperature variations, and even modeling errors. Principal Component Analysis was also used. In [24], the similar methodology was repeated on a beam per impact hammer excitations using the demonstration in [25], focusing on different damage conditions via variations on frequency response functions. The procedure was also utilized on a truss bridge model and found to be successful in the presence of various uncertainties and changing external conditions.

On another identification study on a steel structure, the relationship learning and dimensionality reduction was used with the autoencoders in [26] where the correlation between the stiffness values and modal characteristics were investigated through pattern recognition. Modal parameters were the input information while the resulting damage was the output information of the system. A pre-training was applied on the hidden layers. The procedure was validated numerically and experimentally; also, the performance was reported to be better than the traditional ANNs.

3 Use of CNNs in Damage Detection of Civil Infrastructure

3.1 2D CNNs

Convolutional Neural Networks (CNNs) are a class of supervised ANNs and primarily used in face/object recognition in computers [27, 28]. They have quickly become well known and widely used among the rest of the DL algorithms because of their efficient and fast operation directly on the raw signals [29–33]. When the researchers realized that CNNs are surpassing performance of standard ML procedures in speed and accuracy, more focus has been placed on the application of CNNs on various engineering applications [28, 34, 35]. CNNs are indeed multi-layer feed-forward artificial neural networks in supervised format, inspired by the way mammalian brain's vision cortex operates [36]. CNNs are found advantageous to other networks because of their capability to comprise extraction and classification steps in one body while they process the training on the raw signal. They can adapt to different size of input meanwhile they operate well with the transformations like distortion, skewing, scaling, and translation.

The number of fully connected layers in CNNs typically supersedes the number of layers for pooling and convolution. The CNN architectures are governed by various combinations of subsampling factors: size of kernels; numbers of convolution, fully connected and pooling layers (hyperparameters); and neuron numbers in each of these layers. CNN trainings are in supervised format via the back-propagation (BP) loop procedures. With every BP loop, the sensitivities of network parameters are calculated to improve the parameters until a termination protocol is met. Further information on this procedure is reported in [37–40].

One example study for CNN use for damage identification is a 5-DOF laboratory benchmark structure which is investigated numerically in [41]. The structure demonstrated in [42] was employed for numerical investigation. 2D CNNs were used; therefore, 1D vibration signals were mapped into 2D format to be able to process the network model. The proposed CNN methodology was verified with the acceleration data of El Centro earthquake applied on the structural frame even in the presence of noise. The results were reported to outperform some of the traditional ML-based methods.

In another relevant study [43], a methodology based on CNNs was introduced for identifying and quantifying damage on a concrete bridge. Experimental verification was shown for the presented method where four damage scenarios were studied utilizing the acceleration response of the bridge. 2D CNNs were used again and based on 48 shake table tests, 40 sets of data were assigned for CNN training and 8 sets of data were assigned for validation. 2D CNN performance was reported to be successful for detection and quantification of damage.

In [44, 45], a CNN-based methodology was introduced for damage identification by processing images via functions of transmissibility where the stiffness reductions on members were simulated as damage. A beam and a spring-mass assembly was used to generate data and validate the proposed method using finite element models. The damage detection method was reported to perform successfully.

3.2 1D CNNs

The CNN applications discussed above were based on conventional 2D CNNs which—as the name implies—are a better fit for 2D inputs like videos and images. For 1D signals, One-dimensional CNNs were proposed in [38] for detection of arrhythmia in electrocardiograms and then utilized in numerous applications that employ 1D signals [36, 46–51]. 1D CNNs do have some architectural changes when compared to the 2D counterparts, but in a sense, they can be considered as similar. The performance of 1D CNNs, however, is found to be surpassing 2D CNNs in many platforms [52–55] due to several reasons. The most important one is the fact that 1D CNNs carry much less computational complexity than 2D CNNs. The other reason is 1D CNN architecture is compact and it is designed in a way that even when the training data is not plenty, difficult problems are successfully solved with the smaller number of neurons and hidden layers. On the other hand, for 2D CNNs, when the data is not plenty, overfitting problems tend to arise. On another note, back-propagation and forward-propagation procedures of 2D CNNs require special hardware like GPU farms, while 1D CNNs are mostly fine with standard CPU use on a basic personal computer. All these points make 1D CNNs a better fit for real-time use than 2D CNNs.

A relatively large laboratory structure was used in [40] where 1D CNNs were deployed to train and test 31 damage scenarios. A sensor is placed at each node of the structure and trained separately for that node, where the assigned 1D CNN processed the data collected at the corresponding sensor. The methodology was reported to be successful not only for single but also for double damage scenarios. It was also noted that the damage identification, localization, and quantification were performed in real time. The source code, the test data, benchmark dataset, and accompanied files are shared online on a public website [56]. The benchmark dataset is also published as a conference paper [57]. More information on the large laboratory structure can be found in [58]. Considering the fact that the introduced "damage" was only loosening the bolts at joint locations (an almost negligible rotational stiffness change), it can be stated that the compact 1D CNNs are capable of distinguishing very complex acceleration data with uncorrelated content. It is important to note that a basic personal computer was used in [40], and all of the damage scenarios were predicted accurately, faster than the requirements accepted for real time.

The material presented in [40] was followed by additional experimental studies such as [50] and [32] where the 1D CNN-based approach introduced in [40] was implemented on a Wireless Sensor Network working directly on the data collected by wireless sensors. 1D CNN architecture was in compact form with 2 CNN layers (four neurons/layer) and 2 MLP layers (five neurons/layer). The damage detection and localization performance results were reported to be satisfactory for all damage scenarios imposed on the laboratory frame.

Yet, it was noted that the classifier training operation discussed in [40] and [50] involves long data recordings, which was on the manageable side for a laboratory frame, yet it can be cumbersome (and sometimes unrealistic) for large civil infrastructure. In an attempt to address this, an updated version, "adaptive 1D CNNs" were proposed in [49] and [51] utilizing the data presented in [59]. It was reported that the training was conducted with less effort with the adaptive 1D CNNs, and damage detection, localization, and quantification tasks were processed successfully for all damage conditions discussed in [59].

4 Conclusions and Recommendations for Future Work

As discussed so far, the researchers all over the world have been using DL procedures for detecting, quantifying, and locating damage on civil infrastructure. It has been reported that the CNN procedures are relatively easy to implement since the CNN procedures do not mandate manual extraction of features, which means the users can operate straight with the raw signals. There is also no need for pre-processing because the system can operate directly on unprocessed signal. When it comes to a comparison between 2D and 1D CNNs, it was reported that training for 1D CNNs is a relatively easier process since they are in compact form. On another note, 1D CNNs were reported to comprise lower computational complexity than 2D CNNs. Especially in cases of sparse data, 1D CNNs were still able to perform satisfactorily for detecting, locating, and quantifying damage.

Yet, it is important to note that some of the identification methods per DL are involving supervised algorithms. For large civil engineering infrastructure, it is not always feasible to manage this since obtaining "before damage" and "after damage" recordings is logistically difficult and sometimes even the attempt itself is unrealistic. Based on this, there is a need for development of new methodologies to eliminate the dependency on data, especially for "after damage" conditions. In order to go around the need for data for damaged conditions, one alternative way would be researching the available data for the existing damaged conditions of structures (forming a library of structures) and establishing a link between an undamaged structure and computer simulations of a library of damaged structures. As a final note, it is observed that the number of studies involving unsupervised or semi-supervised studies for damage identification is relatively low; therefore, additional research is required on unsupervised methods.

References

1. Dwivedi, S.K., Vishwakarma, M., Soni, P.A.: Advances and researches on non destructive testing: a review. Mater. Today Proc. **5**(2), 3690–3698 (2018). https://doi.org/10.1016/j.matpr.2017.11.620
2. Wu, X., Ghaboussi, J., Garrett, J.H.: Use of neural networks in detection of structural damage. Comput. Struct. (1992). https://doi.org/10.1016/0045-7949(92)90132-J
3. Frangopol, D.M., Liu, M.: Maintenance and management of civil infrastructure based on condition, safety, optimization, and life-cycle cost. Struct. Infrastruct. Eng. (2007). https://doi.org/10.1080/15732470500253164
4. Ngoan, D.T., Mustafa, G., Osama, A., Onur, A.: Stadium vibration assessment for serviceability considering the vibration duration. In: Proceedings, Annu. Conf. - Can. Soc. Civ. Eng. (2017)
5. Celik, O., Catbas, F.N., Do, N.T., Gul, M., Abdeljaber, O., Younis, A., Avci, O.: Issues, codes and basic studies for stadium dynamics. In: Proc. Second Int. Conf. Infrastruct. Manag. Assess. Rehabil. Tech., Sharjah, UAE (2016)
6. Abdeljaber, O., Hussein, M.F.M., Avci, O.: In-service video-vibration monitoring for identification of walking patterns in an office floor. In: 25th Int. Congr. Sound Vib. Hiroshima, Japan (2018)
7. Chaabane, M., Ben Hamida, A., Mansouri, M., Nounou, H.N., Avci, O.: Damage detection using enhanced multivariate statistical process control technique. In: 2016 17th Int. Conf. Sci. Tech. Autom. Control Comput. Eng. STA 2016 - Proc (2017). https://doi.org/10.1109/STA.2016.7952052
8. Abdeljaber, O., Hussein, M., Avci, O., Davis, B., Reynolds, P.: A novel video-vibration monitoring system for walking pattern identification on floors. Adv. Eng. Softw. (2020). https://doi.org/10.1016/j.advengsoft.2019.102710
9. Mansouri, M., Avci, O., Nounou, H., Nounou, M.: A comparative assessment of nonlinear state estimation methods for structural health monitoring. Conf. Proc. Soc. Exp. Mech. Ser. (2015). https://doi.org/10.1007/978-3-319-15224-0_5
10. Mansouri, M., Avci, O., Nounou, H., Nounou, M.: Iterated square root unscented Kalman filter for state estimation - CSTR model. In: 12th Int. Multi-Conference Syst. Signals Devices, SSD 2015 (2015). https://doi.org/10.1109/SSD.2015.7348243
11. Mansouri, M., Avci, O., Nounou, H., Nounou, M.: Iterated square root unscented Kalman filter for nonlinear states and parameters estimation: three DOF damped system. J. Civ. Struct. Health Monit. **5** (2015). https://doi.org/10.1007/s13349-015-0134-7
12. Avci, O.: Effects of Bottom Chord Extensions on the Static and Dynamic Performance of Steel Joist Supported Floors. Virginia Polytechnic Institute and State University, Virginia (2005)
13. Avci, O., Davis, B.: A Study on Effective Mass of One Way Joist Supported Systems Struct. Congr. 2015 – Proc. (2015). https://doi.org/10.1061/9780784479117.073
14. Avci, O.: Retrofitting Steel Joist Supported Footbridges for Improved Vibration Response Struct. Congr. 2012 - Proc. (2012). https://doi.org/10.1061/9780784412367.041
15. Avci, O., Bhargava, A., Nikitas, N., Inman, D.J.: Vibration annoyance assessment of train induced excitations from tunnels embedded in rock. Sci. Total Environ. (2020). https://doi.org/10.1016/j.scitotenv.2019.134528
16. Avci, O., Setareh, M., Murray, T.M.: Vibration Testing of Joist Supported Footbridges Struct. Congr. 2010 (2010). https://doi.org/10.1061/41130(369)80
17. Avci, O., Bhargava, A., Nikitas, N., Inman, D.J.: Vibrations Assessment of Existing Building Foundations Due to Moving Trains in Underground Tunnels Conf. Proc. Soc. Exp. Mech. Ser. (2021). https://doi.org/10.1007/978-3-030-47634-2_8
18. Morgenthal, G., Hallermann, N.: Quality assessment of Unmanned Aerial Vehicle (UAV) based visual inspection of structures. Adv. Struct. Eng. (2014). https://doi.org/10.1260/1369-4332.17.3.289

19. Ghahramani, Z.: Probabilistic machine learning and artificial intelligence. Nature. (2015). https://doi.org/10.1038/nature14541
20. Kwon, D., Kim, H., Kim, J., Suh, S.C., Kim, I., Kim, K.J.: A survey of deep learning-based network anomaly detection. Cluster Comput. (2017). https://doi.org/10.1007/s10586-017-1117-8
21. Hinton, G.E., Salakhutdinov, R.R.: Reducing the dimensionality of data with neural networks. Science. **313** (2006). https://doi.org/10.1126/science.1127647
22. Patterson, J., Gibson, A.: Deep Learning: A Practitioner's Approach. O'Reilly Media, Newton, MA (2017). https://doi.org/10.1038/nature14539
23. Fallahian, M., Khoshnoudian, F., Meruane, V.: Ensemble classification method for structural damage assessment under varying temperature. Struct. Health Monit. (2017). https://doi.org/10.1177/1475921717717311
24. Fallahian, M., Khoshnoudian, F., Talaei, S., Meruane, V., Shadan, F.: Experimental validation of a deep neural network—sparse representation classification ensemble method. Struct. Des. Tall Spec. Build. (2018). https://doi.org/10.1002/tal.1504
25. Shadan, F., Khoshnoudian, F., Esfandiari, A.: A frequency response-based structural damage identification using model updating method. Struct. Control Health Monit. (2016). https://doi.org/10.1002/stc.1768
26. Pathirage, C.S.N., Li, J., Li, L., Hao, H., Liu, W., Ni, P.: Structural damage identification based on autoencoder neural networks and deep learning. Eng. Struct. (2018). https://doi.org/10.1016/j.engstruct.2018.05.109
27. Krizhevsky, A., Sutskever, I., Hinton, G.E.: ImageNet classification with deep convolutional neural networks. Adv. Neural Inf. Process. Syst. **2012**, 1097–1105 (2012). https://doi.org/10.1145/3065386
28. Kiranyaz, S., Waris, M.A., Ahmad, I., Hamila, R., Gabbouj, M.: Face segmentation in thumbnail images by data-adaptive convolutional segmentation networks. In: 2016 IEEE Int. Conf. Image Process., pp. 2306–2310 (2016). https://doi.org/10.1109/ICIP.2016.7532770
29. Kiranyaz, S., Avci, O., Abdeljaber, O., Ince, T., Gabbouj, M., Inman, D.J.: 1D convolutional neural networks and applications: a survey. Mech. Syst. Signal Process. **151** (2021). https://doi.org/10.1016/j.ymssp.2020.107398
30. Avci, O., Abdeljaber, O., Kiranyaz, S., Hussein, M., Gabbouj, M., Inman, D.J.: A review of vibration-based damage detection in civil structures: from traditional methods to machine learning and deep learning applications. Mech. Syst. Signal Process. (2021). https://doi.org/10.1016/j.ymssp.2020.107077
31. O. Avci, O. Abdeljaber, S. Kiranyaz, S. Sassi, A. Ibrahim, M. Gabbouj, One Dimensional Convolutional Neural Networks for Real-Time Damage Detection of Rotating Machinery, Conf. Proc. Soc. Exp. Mech. Ser., 2021
32. Avci, O., Abdeljaber, O., Kiranyaz, S., Inman, D.: Convolutional Neural Networks for Real-Time and Wireless Damage Detection, Conf. Proc. Soc. Exp. Mech. Ser. (2020). https://doi.org/10.1007/978-3-030-12115-0_17
33. O. Avci, O. Abdeljaber, S. Kiranyaz, Structural Damage Detection in Civil Engineering with Machine-Learning: Current State of the Art, Conf. Proc. Soc. Exp. Mech. Ser., 2021
34. Ciresan, D.C., Meier, U., Gambardella, L.M., Schmidhuber, J.: Deep, big, simple neural nets for handwritten digit recognition. Neural Comput. **22**, 3207–3220 (2010). https://doi.org/10.1162/NECO_a_00052
35. Scherer, D., Müller, A., Behnke, S.: Evaluation of pooling operations in convolutional architectures for object recognition. In: Proc. 20th Int. Conf. Artif. Neural Networks Part III, pp. 92–101. Springer, Berlin (2010). https://doi.org/10.1007/978-3-642-15825-4_10
36. Kiranyaz, S., Ince, T., Gabbouj, M.: Personalized monitoring and advance warning system for cardiac arrhythmias. Sci. Rep. **7** (2017). https://doi.org/10.1038/s41598-017-09544-z
37. Kiranyaz, S., Ince, T., Hamila, R., Gabbouj, M.: Convolutional neural networks for patient-specific ECG classification. In: Proc. Annu. Int. Conf. IEEE Eng. Med. Biol. Soc. EMBS (2015). https://doi.org/10.1109/EMBC.2015.7318926
38. Kiranyaz, S., Ince, T., Gabbouj, M.: Real-time patient-specific ECG classification by 1-D convolutional neural networks. IEEE Trans. Biomed. Eng. **63**, 664–675 (2016). https://doi.org/10.1109/TBME.2015.2468589
39. Kiranyaz, S., Ince, T., Abdeljaber, O., Avci, O., Gabbouj, M.: 1-D convolutional neural networks for signal processing applications. In: ICASSP, IEEE Int. Conf. Acoust. Speech Signal Process. - Proc. (2019). https://doi.org/10.1109/ICASSP.2019.8682194
40. Abdeljaber, O., Avci, O., Kiranyaz, S., Gabbouj, M., Inman, D.J.: Real-time vibration-based structural damage detection using one-dimensional convolutional neural networks. J. Sound Vib. **388**, 154–170 (2017). https://doi.org/10.1016/j.jsv.2016.10.043
41. Yu, Y., Wang, C., Gu, X., Li, J.: A novel deep learning-based method for damage identification of smart building structures. Struct. Health Monit. **18**, 143–163 (2019). https://doi.org/10.1177/1475921718804132
42. Wu, Y.M., Samali, B.: Shake table testing of a base isolated model. Eng. Struct. (2002). https://doi.org/10.1016/S0141-0296(02)00054-8
43. Khodabandehlou, H., Pekcan, G., Fadali, M.S.: Vibration-based structural condition assessment using convolution neural networks. Struct. Control Health Monit. (2018). https://doi.org/10.1002/stc.2308
44. Cofre-Martel, S., Kobrich, P., Lopez Droguett, E., Meruane, V.: Deep convolutional neural network-based structural damage localization and quantification using transmissibility data. Shock Vib. (2019). https://doi.org/10.1155/2019/9859281
45. Cofré, S., Kobrich, P., López Droguett, E., Meruane, V.: Transmissibility based structural assessment using deep convolutional neural network. In: Proc. ISMA 2018 - Int. Conf. Noise Vib. Eng. USD 2018 - Int. Conf. Uncertain. Struct. Dyn. (2018)
46. Kiranyaz, S., Gastli, A., Ben-Brahim, L., Alemadi, N., Gabbouj, M.: Real-time fault detection and identification for MMC using 1D convolutional neural networks. IEEE Trans. Ind. Electron. (2018). https://doi.org/10.1109/TIE.2018.2833045
47. Ince, T., Kiranyaz, S., Eren, L., Askar, M., Gabbouj, M.: Real-time motor fault detection by 1-D convolutional neural networks. IEEE Trans. Ind. Electron. (2016). https://doi.org/10.1109/TIE.2016.2582729
48. Avci, O., Abdeljaber, O., Kiranyaz, S., Inman, D.: Structural damage detection in real time: implementation of 1D convolutional neural networks for SHM applications. In: Niezrecki, C. (ed.) Struct. Heal. Monit. Damage Detect Proc. 35th IMAC, A Conf. Expo. Struct. Dyn. 2017, vol. 7, pp. 49–54. Springer International Publishing, Cham (2017). https://doi.org/10.1007/978-3-319-54109-9_6
49. Abdeljaber, O., Avci, O., Kiranyaz, M.S., Boashash, B., Sodano, H., Inman, D.J.: 1-D CNNs for structural damage detection: verification on a structural health monitoring benchmark data. Neurocomputing. (2017). https://doi.org/10.1016/j.neucom.2017.09.069
50. Avci, O., Abdeljaber, O., Kiranyaz, S., Hussein, M., Inman, D.J.: Wireless and real-time structural damage detection: a novel decentralized method for wireless sensor networks. J. Sound Vib. (2018)
51. Avci, O., Abdeljaber, O., Kiranyaz, S., Boashash, B., Sodano, H., Inman, D.J.: Efficiency validation of one dimensional convolutional neural networks for structural damage detection using a SHM benchmark data. In: 25th Int. Congr. Sound Vib. (2018)

52. Eren, L.: Bearing fault detection by one-dimensional convolutional neural networks. Math. Probl. Eng. (2017). https://doi.org/10.1155/2017/8617315
53. Abdeljaber, O., Sassi, S., Avci, O., Kiranyaz, S., Abulrahman, I., Gabbouj, M.: Fault detection and severity identification of ball bearings by online condition monitoring. IEEE Trans. Ind. Electron. (2018) https://ieeexplore.ieee.org/document/8584489
54. Li, D., Zhang, J., Zhang, Q., Wei, X.: Classification of ECG signals based on 1D convolution neural network. In: 2017 IEEE 19th Int. Conf. e-Health Networking, Appl. Serv. Heal. 2017 (2017). https://doi.org/10.1109/HealthCom.2017.8210784
55. Xiong, Z., Stiles, M., Zhao, J.: Robust ECG signal classification for the detection of atrial fibrillation using novel neural networks. In: 2017 Comput. Cardiol. Conf. (2018). https://doi.org/10.22489/cinc.2017.066-138
56. Avci, O., Kiranyaz, S., Abdeljaber, O.: StructuralDamageDetection.com (Public Website). http://www.structuraldamagedetection.com/ (2019)
57. O. Avci, O. Abdeljaber, S. Kiranyaz, M. Hussein, M. Gabbouj, D.J. Inman, A New Benchmark Problem for Structural Damage Detection: Bolt Loosening Tests on a Large-Scale Laboratory Structure, Conf. Proc. Soc. Exp. Mech. Ser., 2021
58. Abdeljaber, O., Younis, A., Avci, O., Catbas, N., Gul, M., Celik, O., Zhang, H.: Dynamic testing of a laboratory stadium structure. Geotech. Struct. Eng. Congr. **2016**, 1719–1728 (2016). https://doi.org/10.1061/9780784479742.147
59. Dyke, S., Bernal, D., Beck, J., Ventura, C.: Experimental phase II of the structural health monitoring benchmark problem. In: Proc. 16th ASCE Eng. Mech. Conf., pp. 1–7 (2003)

Transfer Learning from Audio Domains a Valuable Tool for Structural Health Monitoring

Eleonora M. Tronci, Homayoon Beigi, Maria Q. Feng, and Raimondo Betti

Abstract Today, the application of artificial neural network tools to define models that mimic the dynamic behavior of structural systems is a wide-spread approach. A fundamental issue in developing these strategies for damage assessment in civil structures is represented by the unbalanced nature of the available databases, which commonly contain plenty of data coming from the structure under healthy operational conditions and very few samples from the system in unhealthy conditions since the structure would have failed by that time. Consequently, the learning task, carried on with standard deep learning approaches, becomes case-dependent and tends to be specialized for a particular structure and for a very limited number of damage scenarios.

This work presents a framework for damage classification in structural systems intended to overcome such limitations. In this methodology, the model is trained to gain knowledge in the learning task from a rich acoustic dataset (source domain), acquiring higher-level features characterizing vibration traits from a rich acoustic dataset. This knowledge is then transferred to a target domain, with much less training data, such as a structural system, in order to classify its structural condition.

The framework starts with constructing a time-delay neural network (TDNN) structure, trained on the VoxCeleb dataset, in the speech domain. The input of the network consists of Cepstral and pitch features extracted from the audio records. Higher-level features, the x-vectors, speaker embeddings, capturing neural outputs of the network's intermediate layers, are derived and then used to train a probabilistic linear discriminant analysis (PLDA) model to provide a probabilistic discriminant model for speaker comparison. These features collect generic information regarding the source domain and characterize a classification process based on the frequency content of signals, which is not strictly dependent on the original acoustic domain. Because of the non-case-dependent nature of the x-vector embeddings (features), they can be used to train an alternative PLDA model to address a damage classification task, considering vibration measurements coming from a different system, a structural one which represents the target domain. The simulated data from the 12 degrees of freedom benchmark shear-building structure provided by the IASC-ASCE Structural Health Monitoring Group are studied to verify the proposed framework's effectiveness.

Keywords Transfer learning · Structural Health Monitoring · Mel frequency cepstral coefficients · Time-delay neural network · x-Vector features

1 Introduction

The aim of vibration-based Structural Health Monitoring (SHM) methods is to implement a strategy to correctly detect damage by assessing changes in the identified vibration response of civil structures [1]. Identifying structural damage scenarios through information extracted from the structure's dynamic response can be addressed from different perspectives. An efficient way to do so is to apply Pattern Recognition and Machine Learning (PRML) strategies [1] that focus on detecting meaningful patterns in the features, defined as Damage Sensitive Features (DSFs), representing the conditions of the structural system.

E. M. Tronci (✉) · M. Q. Feng · R. Betti
Department of Civil Engineering and Engineering Mechanics, Columbia University, New York, NY, USA
e-mail: et2501@columbia.edu

H. Beigi
Recognition Technologies, Inc., South Salem, NY, USA

K. Grimmelsman (ed.), *Dynamics of Civil Structures, Volume 2*, Conference Proceedings of the Society for Experimental Mechanics Series, https://doi.org/10.1007/978-3-030-77143-0_11

A fundamental issue in the development of data-based PRML strategies for damage assessment is represented by the fact that the datasets available for mechanical and civil applications are quite limited in number and the amount of information provided. Unfortunately, when the aim is to classify between healthy and damaged conditions in a structure, it is extremely rare to have enough data for training the learning algorithm to recognize the unhealthy (or damaged) condition. This lack of information can cause ill-conditioning in training classification and prediction models built using machine learning and deep learning algorithms. Namely, the dataset does not contain sufficient information from all the different classes that need to be identified. Consequently, for mechanical and civil applications, the learning task becomes case-dependent and tends to be specialized for a particular structure and a very limited number of damage scenarios.

To overcome such a limitation imposed by the insufficiency of exhaustive failure datasets for civil and mechanical systems, we propose to take advantage of the richness and knowledge in the learning task gained from a source domain, with extensive and exhaustive datasets, and to transfer that same knowledge to a target domain, with much less information. This operation goes under the name of "transfer learning" and is emerging as a fundamentally new way of thinking in this data-driven and AI-centric era to move data around to enhance knowledge extraction effectively. Transfer Learning (TL) consists of transferring knowledge from one domain to another to take advantage of that knowledge for a different but related task or domain. The reader is referred to [2–4] for an overview of the different transfer learning methods. In this work, a transfer learning method is implemented to gain knowledge from automatic speech recognition, adapting the learning task to various classification tasks in structural domains or datasets. One of the advantages of deep learning is acquiring a hierarchy of feature representations from low-level features to more abstract higher-level features ones [4, 5]. Therefore, pre-training [6–8] on a rich domain to explicitly learn intermediate-level features in the neural network can be useful for several different tasks. The intermediate layers in a neural network's architecture trained on speech data appear to be not specific to any particular task, while the higher layers are task-specific [9]. Supervised training using out-of-domain data is also a form of pre-training, and it has been used to learn multilingual bottleneck features in [10, 11].

The present work focuses on the development of a vibration-based SHM framework for damage classification in civil and mechanical structures (e.g., the target domain) by cleverly tapping from rich prediction and classification models that were trained on large datasets, such as audio data, largely used in speech and speaker recognition problems (the source domain). The common ground between these two domains is that they are both based on vibrating systems. It is possible to learn from a source domain, not directly related to the structural problem but with an enormously rich database, and transfer this knowledge to the target domain, based on the shared underlying physics (on vibrating systems and their time-frequency features). The transfer process is facilitated by the adoption of Mel Frequency Cepstral Coefficients as features [12, 13]. These coefficients are vastly used in speaker recognition tasks, and their use as DSFs in structural damage assessment should facilitate the transition between the audio and structural domains and enable an explainable artificial intelligence.

2 Data

Audio Domain In the following work, the audio records collection adopted as a source domain is the VoxCeleb dataset. It is an audio-visual dataset consisting of short clips of human speech, extracted from interview videos uploaded to YouTube. It consists of records belonging to more than 7000 speakers, where each audio-video segment is at least 3 s long for a total of more than 2000 h of recorded data. Only the audio component of the dataset is considered in this study, while the videos are discarded. The dataset was released in two stages, as VoxCeleb1 and VoxCeleb2 [14]. VoxCeleb1 contains over 100,000 utterances for 1251 celebrities, while VoxCeleb2 contains over 1 million utterances from over 6000 celebrities extracted from videos uploaded to YouTube. The datasets are fairly gender-balanced, (VoxCeleb1—55% male, VoxCeleb2—61% male). The speakers span a wide range of different ethnicities, accents, professions, and ages. The records are characterized by corruption with real-world noise, consisting of background chatter, laughter, overlapping speech, room acoustics, and a range in the quality of recording equipment and channel noise.

Structural Domain The target domain is represented by the dataset obtained from numerical simulations of the 12DOF shear-type building model used as benchmark model from the IASC-ASCE SHM Task Group [15]. The response of the system is simulated under seven (one undamaged and six damaged) structural conditions, considering different levels of damage (from heavy damage) such as no stiffness in any of the braces of the first and third stories (damage case D2) to light damage, e.g., 33% stiffness loss in one brace (damage case D6). The reader is referred to [15] for a more detailed description of the simulated system and to Fig. 1 for the description of the damaged scenarios.

The simulation of the system response is implemented considering 16 sensor locations. The target domain dataset comprises 35,200 and 9600 simulated time-histories for the undamaged and damaged scenarios, respectively (100

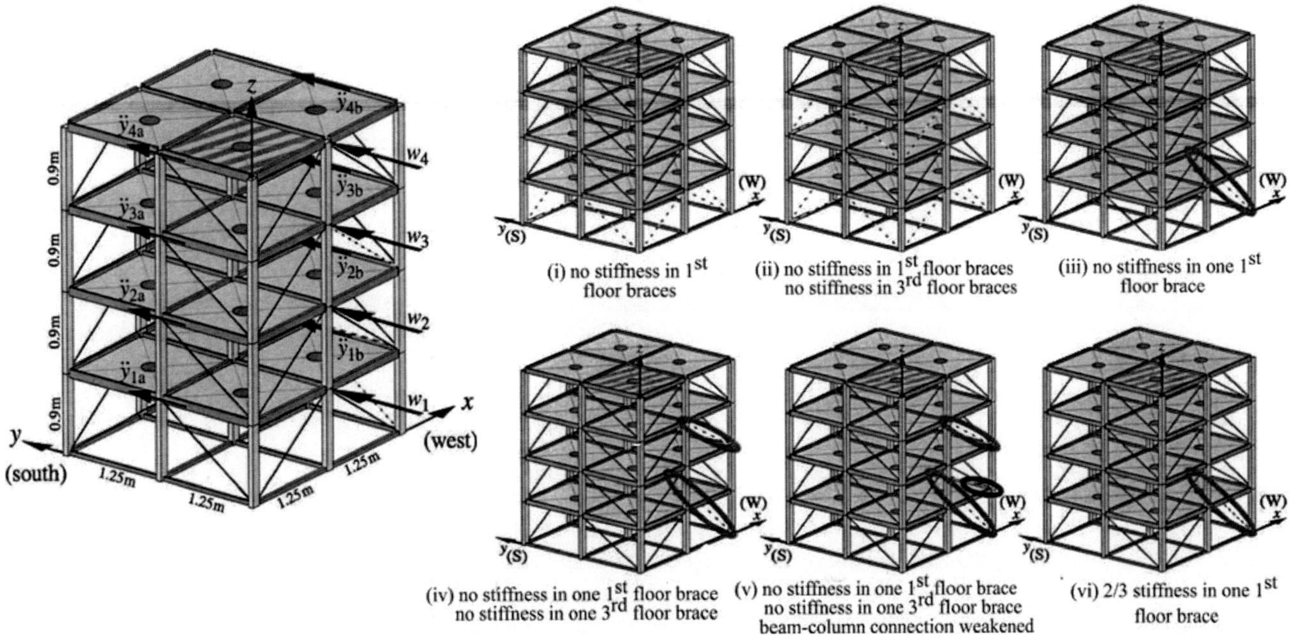

Fig. 1 Diagram of 12DOF analytical model and of the six damage scenarios. The w_i are excitations and the $\ddot{y}_{ij}$ are accelerometer measurements in the y-direction; (the ones in the x-direction are omitted for clarity)

simulations for each damage condition). The disparity in the number of damaged and undamaged realizations (undamaged 78.6%, damaged 21.4%) mimics the unbalanced nature that commonly characterizes the measurement collection for civil and mechanical monitored systems. Each measurement realization is 10 min long and is sampled at 0.0025 s.

3 Analysis

The proposed procedure relies on creating a model trained partially on the audio records of the source domain and partly on the structural vibration response of the structure. The model will learn the higher-level features characterizing vibration records from the rich audio dataset and then specialize its knowledge on the chosen structural dataset. The goal is to enrich the model's ability to discriminate between classes on the audio records, presenting multiple different classes with more information to learn. The same model, trained only on the structural dataset, would be ill-conditioned since these datasets tend to be unbalanced in the number of records available for the different classes and would not provide a robust and reliable collection of information.

Two different types of classifications are addressed in this work. First, a binary classification is carried on, where the final aim is to label the tested records from the structure system as either "damaged" or "undamaged." Then a multiclass classification is exploited where the goal is to distinguish between the undamaged condition and the different types of damage scenarios. The proposed TL strategy consisting of four steps is addressed: 1. the Data Preparation; 2. the Feature Extractions; 3. the Training Phase; 4. the Test Phase.

Data Preparation In the data preparation stage, the source domain and the target domain datasets need to be properly defined and uniformed to avoid ill-conditioned relations between the two domains. In this stage, the train and test datasets are constructed, where the first collects data from the audio dataset and a section from the structural systems, while the test set consists of data only from the structural system.

It is fundamental to highlight how the two target domains associated with structural systems and the source domain present different frequency content. Generally, audio records show a frequency content that can space from low-frequency value up to kHz, and the sampling frequency usually adopted to collect those records is very high (8 kHz or 16 kHz). Lower frequency contents characterize structural systems that rarely exceed the 20–30 Hz range, and consequently, the needed sampling frequency is much smaller (100–200 Hz) with respect to the one used for audio records. This discrepancy in frequency content requires a domain adaptation procedure to uniform the frequency spectrum of the source and target

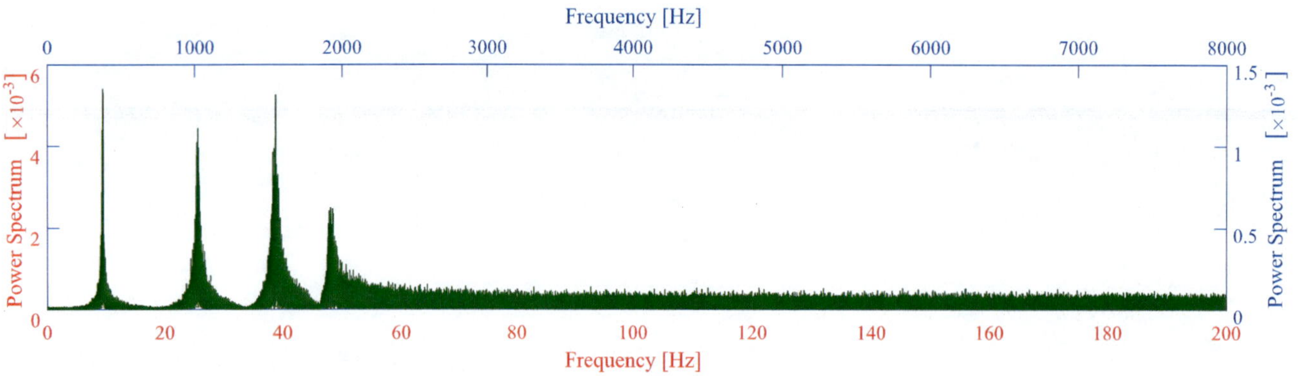

Fig. 2 Power spectrum of an undamaged record for the first sensor. In red the original frequency scale of the structural dataset, in blue the modified scale obtained after frequency domain transformation

domains. In this work, the structural system presents records sampled at 400 Hz, while the VoxCeleb utterances are recorded at 16 kHz. To achieve an equivalence in terms of frequency content between the domains, a transformation procedure is implemented to bring the structural frequency content to the same frequency range of the audio records (Fig. 2). To build a richer audio dataset, both the audio and structural datasets have been augmented. In the VoxCeleb dataset, the records are augmented with reverberation, noise, music, and babble using the MUSAN dataset. Once the corrupted records are created, they are combined with the original clean data so that the trained model can learn and discriminate between cleaned signals and signal with noise and disturbances. For the structural dataset, the original raw signal for the numerical simulation is corrupted with white Gaussian noise characterized by 10% RMS, and in each simulation, a small perturbation on the mass and stiffness of each degree of freedom is introduced.

The dataset adopted in the training phase consists of two groups: the first group collects the audio features to train the first part of the model, and the second group consists of records belonging to the structural system. The VoxCeleb dataset will be used entirely for the training, while for the structural systems, only 80% of the records will be adopted for the training. The remaining 20% of the structural data for the creation of the test set.

Feature Extraction In the present work, the Mel Frequency Cepstral Coefficients [12] are adopted as damage sensitive features. For both the VoxCeleb audio domain and for the structural target domain, which is properly transformed, the MFCC vector consists of 30 coefficients per frame. Besides the Mel Frequency Cepstral Coefficients, another set of features is considered and added into the feature vector used to train the classification model, pitch, delta-pitch, and probability of voicing features. Pitch is a perceived quantity related to the fundamental frequency of vibration of the system to which it is referred to. The final damage sensitive feature vector adopted is a 33-dimensional vector. The extraction process is implemented for every record collected in the audio dataset and structural measurements for the training set, and the structural records collected in the test dataset.

Training Phase The features extracted in the previous step are prepared for the training process. They are randomly selected and assigned to the proper tags of classes, creating the dataset, which will be the input for a neural network model. This process is done for both the structural dataset and the audio set. Then, the cepstral mean normalization procedure is applied to the features to make them all zero-mean and remove the convolved noise within the signal. Additionally, for the audio dataset, the possible silence frames are removed. In [16], the authors show that training a PLDA classifier on fixed-length embeddings extracted from the higher layers of a speaker recognition TDNN (which they refer to as "x-vectors") achieves superior performance on out-of-class speaker recognition. Ananthram et al. [17] successfully implement a similar strategy for automated emotion detection in speech. Following an equivalent approach, the classification task is implemented here, assuming that such a network learns dense representations of speech segments in its upper layers and that these abstract information can be used to classify later the structural health condition of the 12DOF system. The features are derived from the audio domain and used to pre-train a TDNN architecture on a speaker recognition task [12]. Here, the same 9-layer architecture and training methodology adopted in [16] is implemented, using the training script published as part of the Kaldi toolkit [18]. Time-Delay Neural Network is a multilayer artificial neural network architecture able to capture an unknown system's dynamics by modeling a flexible-structured network that will imitate the system by adaptively changing its parameters. This architecture maps a finite time sequence into a single output. Each layer of a TDNN processes a context window from the previous layer, which means that lower layers will have a smaller receptive field and therefore model local features, and higher layers will have a bigger receptive field and thus model long-term dependencies from the slice of features. The input features are 33-dimensional filterbanks with a frame-length of 25 ms, mean-normalized over a sliding window of up to 3 s.

The TDNN is trained to classify the N classes in the training data. In the audio features, those N classes represent the speakers. In the structural monitoring case, those classes will be two for the binary classification (damaged and undamaged) or N representing different damage classes for the multiclass classification task. In a second stage of the training process, once the presented TDNN architecture is trained using the VoxCeleb audio dataset, it is possible to extract the higher-level features, the x-vectors embedded in the network's intermediate layers. Out of the 9th layer network, the 6th layer is extracted as an x-vector, where the basic information that makes a speaker is learned. These features collect generic information regarding the source domain and characterize a classification process based on the frequency content of signals, which is not strictly dependent on the audio domain. After extracting the x-vector, the mean feature vector is computed, and it is removed for centering the evaluation x-vectors. Starting from the feature x-vector, the trained ability of the model to separate classes is maximized in the final stage in our pipeline with the application of an LDA/PLDA [16, 19] strategy. First, the feature vector is projected into a lower-dimensional G features space adopting the Linear Discriminant Analysis approach, which identifies the subspace where the data between different classes is most spread out, relative to the spread within each category. Then, to build an accurate probability model for the new unknown class and find the likelihood estimation, a PLDA model is trained, which takes that projected space derived in the LDA step and rotate it and shift it in a way that stays in the G dimensional space. The PLDA model assumes based on the prior probabilities associated with the LDA analysis and sets an actual probability distribution.

Test Phase In the test phase, the model's input is represented by the features extracted from the test set, and the output is the log-likelihood ratio or similarity score, s, given for each analyzed record and for each class, which says how close each record is to one of the classes. The record is assigned to the class with the higher score. The evaluation data protocol comprises a list of trials, each corresponding to a pair of structural records. The accuracy of the classification procedures is evaluated using the same verification approach adopted for the speaker recognition task, where in the damage detection, it is verified if the tested record matches or not the target health condition. At the end, the performance of the classification model will be given in terms of the Equal Error Rate (EER) or in terms of accuracy. Equal error rate is used to determine the threshold value for a system when its false acceptance rate (FAR) and false rejection rate (FRR) are equal. The accuracy is computed instead as the ration between the total number of true positive and true negative cases and the total number of inspected cases. This rate is then known as the Equal Error Rate. The graphic representation of the miss probability by the false alarm rate is Detection Error Trade-Off (DET) Curve [12].

4 Results

The results are addressed in Table 1 in terms of EER and accuracy for the two classification tasks: a binary classification between undamaged and damaged conditions; a multiclass classification to differentiate the classes according to the damage type. The two tasks are carried on considering two different configurations of the trained model on the VoxCeleb dataset. In the first arrangement, the dataset presents no noise augmentation, while in the second, the MUSAN dataset is adopted to augment the audio dataset.

Initially, the damage detection framework is tested on a dataset (Case A) with well separated classes which collects the data from the undamaged and the first two damage conditions (D1 and D2). These two damage conditions represent the most severe damage conditions and contemplate, respectively, no stiffness in the braces of the first story (D1) and no stiffness in any of the braces of the first and third stories (D2). Subsequently, the framework performance is tested considering the dataset with the undamaged and all the damaged conditions, from D1 to D6, included (Case B). The performance of the two classification tasks is visualized in Figs. 3 and 5, given in terms of the DET curve.

Audio Multiclass Classification: Speaker Recognition Task

The preliminary results coming for the implementation of the presented strategy seem to be extremely promising. The first step concerns the investigation of the accuracy of the TDNN model trained on the VoxCeleb dataset. It is important to check this model's ability to achieve the speaker classification task using the previously addressed features, parameters, and structure. The test results show the robustness of the model trained on the source domain in the multiclass classification, with an EER of 3.259%. Additionally, the noise augmentation of the dataset with the MUSAN disturbances shows a beneficial effect in the classification task, leading to an EER of 2.641%. The obtained results showed an improvement in adopting the pitch features and the MFCCs with respect to the original reference formulation of the framework, based only on the MFCCs, which lead to an EER of 3.128% considering the augmented dataset.

Table 1 Results for the different classification tasks considering the VoxCeleb dataset with and without the noise augmentation with the MUSAN dataset for case A and case B

Task	VoxCeleb unaugmented		VoxCeleb noise augmented	
	EER [%]	Accuracy [%]	EER [%]	Accuracy [%]
Binary classification (case A)	3.08	97.15	3.08	96.77
Multiclass classification (case A)	3.85	96.02	4.23	95.64
Binary classification (case B)	23.13	76.90	21.06	78.99
Multiclass classification (Case B)	18.32	81.44	17.45	82.47

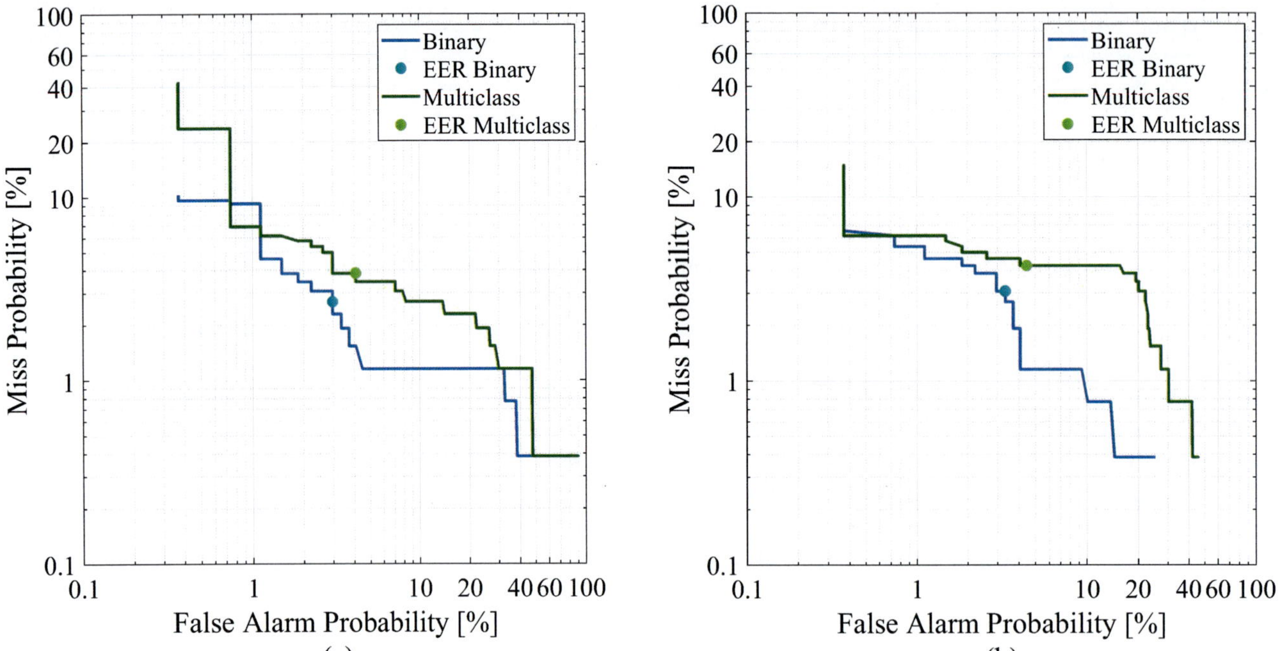

Fig. 3 DET curves for the binary and multiclass tasks, considering the dataset in case A: (**a**) unaugmented VoxCeleb dataset; (**b**) noise augmented VoxCeleb dataset

Structural Binary and Multiclass Classification for Case A

The model trained on the dataset for case A performs with a high accuracy both in the binary and multiclass classification tasks. In this configuration only the undamaged and first two damaged conditions are considered in the training. In the binary classification (Fig. 3) the model achieves an EER of 3.08% (LDA dimension = 3) considering both the VoxCeleb dataset without noise augmentation and its augmented version. In the multiclass classification (Fig. 3) the model achieves an EER of 3.85% and 4.23% (LDA dimension = 3) considering, respectively, the VoxCeleb dataset without noise augmentation and its augmented version. It is evident, how in these classification tasks, associated with the dataset in case A, the addition of noise in the audio domain does not lead to a better performance.

Figure 4 presents the LDA-transformed x-vectors for the three LDA dimensions in the augmented VoxCeleb dataset. As expected the damaged and undamaged scenarios are correctly separated in the binary classification, and even when the discrimination gets more granular in the multiclass case, the two damage classes are correctly classified independently. It is evident how the first and second LDA dimensions, which are associated with the biggest eigenvalues, play a key role in separating the x-vectors features related to the two classes.

Structural Binary and Multiclass Classification for Case B

The binary classification task implemented for the dataset in case B leads to an EER of 23.13% (LDA dimension = 5) when considering the VoxCeleb dataset without noise augmentation and to an EER of 21.06% (LDA dimension = 6) when the MUSAN dataset is considered for augmenting the dataset (Fig. 5). The results demonstrate the capability achievable in the classification task by training the TDNN architecture on the audio source. Figure 6 presents the LDA-transformed

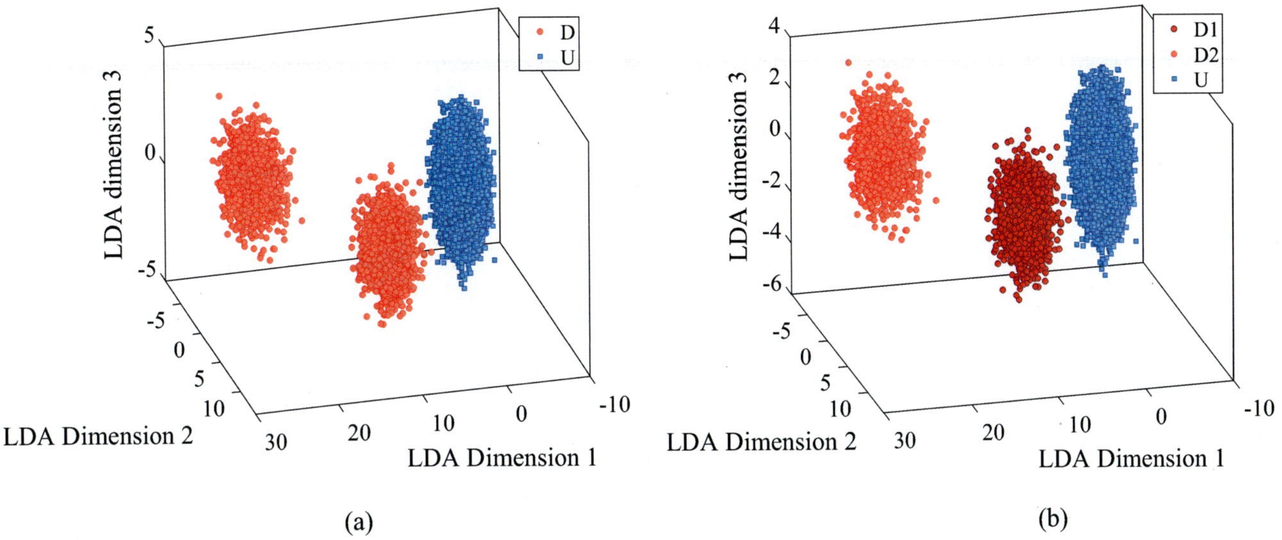

Fig. 4 Three components of x-vectors transformed via LDA for the noise augmented VoxCeleb dataset, considering the dataset in case A: (**a**) binary classification; (**b**) multiclass classification

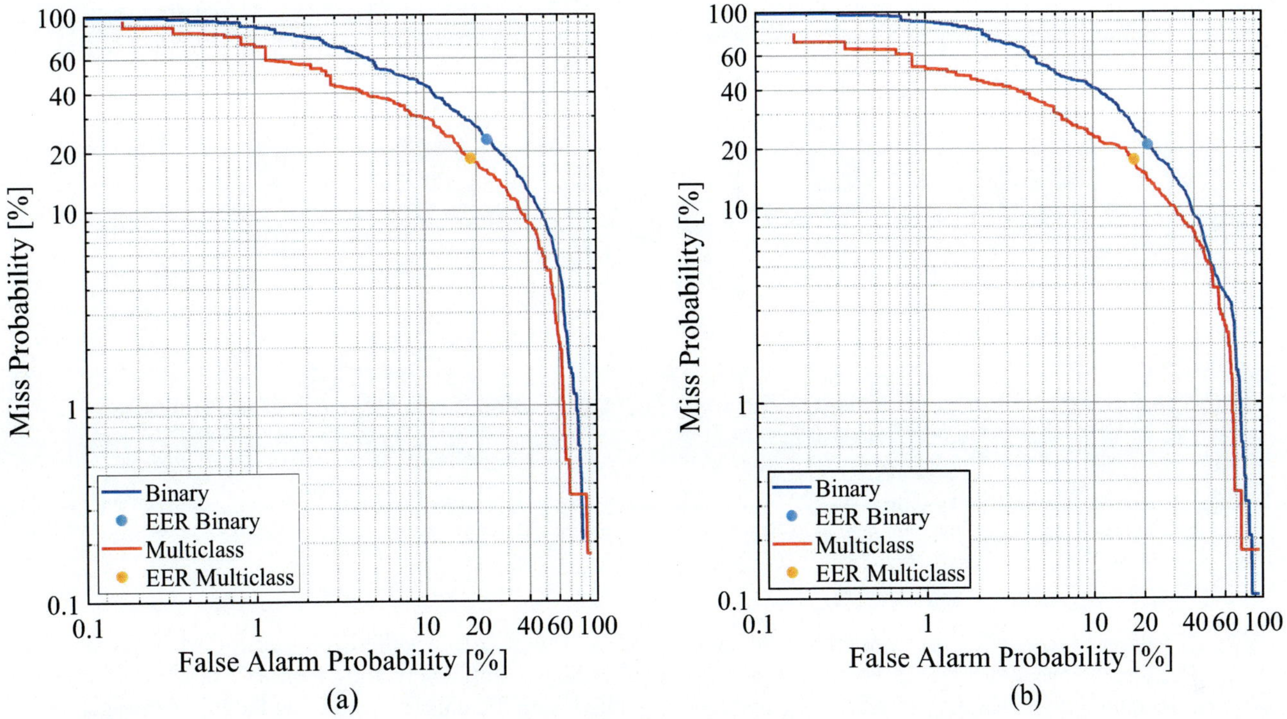

Fig. 5 DET curves for the binary and multiclass tasks, considering the dataset in case B: (**a**) unaugmented VoxCeleb dataset. (**b**) Noise augmented VoxCeleb dataset

x-vectors for the first four LDA dimensions in the augmented VoxCeleb dataset. The majority of the cases are correctly classified. However, some of the conditions, belonging to the less severe damage cases (i.e., damage scenarios D6), are wrongly classified. In this case, it is evident how the first LDA dimension, which is associated with the biggest eigenvalue, plays a key role in separating the x-vectors features related to the two classes, while the other features have a low influence.

The multiclass classification task implemented for the dataset in case B leads to an EER of 18.32% (LDA dimension = 5) when considering the VoxCeleb dataset without noise augmentation and to an EER of 17.45% (LDA dimension = 3) when

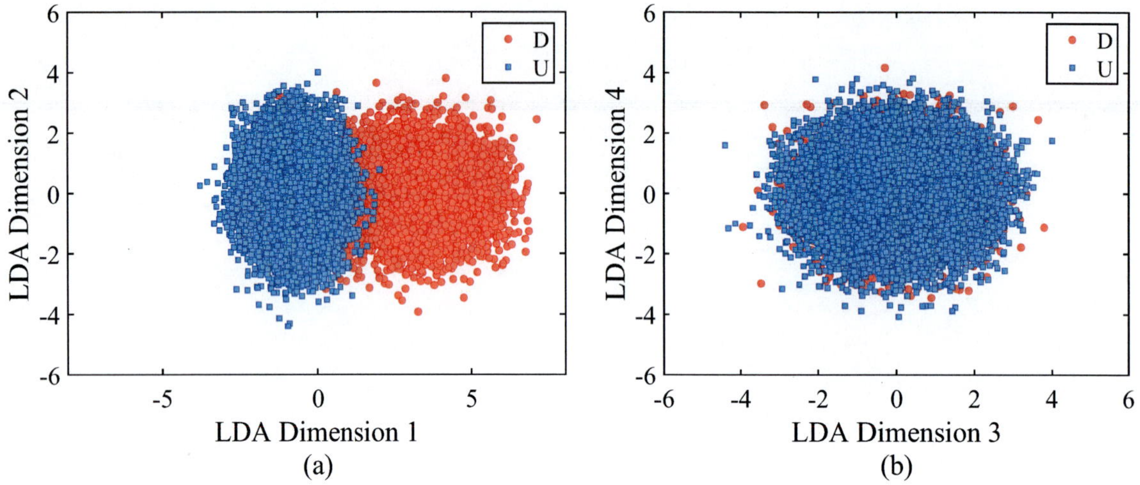

Fig. 6 First four components of x-vectors transformed via LDA for the binary classification task considering the noise augmented VoxCeleb dataset in case B: (**a**) LDA Dimension 1 vs. LDA Dimension 2; (**b**) LDA Dimension 3 vs. LDA Dimension 4

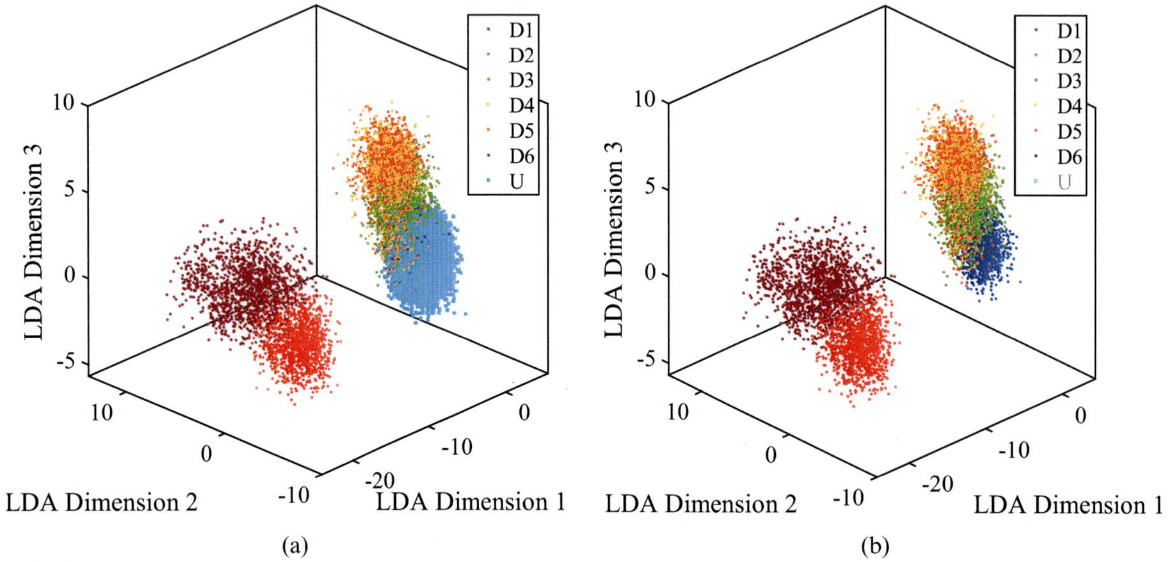

Fig. 7 Three components of x-vectors transformed via LDA for the multiclass by damage type classification task considering the noise augmented VoxCeleb dataset in case B: (**a**) LDA Dimension 1 vs. LDA Dimension 2 for damaged and undamaged scenarios; (**b**) LDA Dimension 1 vs. LDA Dimension 2 for damaged scenarios

the MUSAN dataset is considered for augmenting the dataset (Fig. 5). Figure 7 shows the LDA-transformed x-vectors for the three LDA dimensions in the case of noise augmented VoxCeleb dataset. Intuitively, the x-vectors features linked with the most severe damages, scenarios D1 and D2, are perfectly separated from the clusters describing the other damaged and undamaged scenarios. The clusters representing the features related to damage type D4 and D5, which are less severe failure mechanisms with respect to cases D1 and D2 and consist of more localized damage, are still correctly classified. Damage D3, which still interests the loss of stiffness of a whole structural element, is correctly distinguished from the undamaged conditions, even with some wrongly classified cases. On the other hand, damage type D6, which corresponds to a partial loss of stiffness for one structural element, is improperly classified.

5 Conclusion

The present study addresses a novel detection SHM strategy based on adopting a transfer learning approach from the audio domain to the structural domain. The framework is based on constructing a richer Probabilistic Linear Discriminant Analysis model starting from the x-vector features extracted from a Time-Delay Neural Network model trained on audio features. The trained model is then used to classify the test data from a structural system in a binary and multiclass classification task.

The proposed methodology is presented and tested to assess damage classification in a simulated 12 DOF shear-type system. The results show strong reliability for the trained model obtained by training the TDNN architecture on the audio source.

The model shows an excellent performance in both binary (damaged and undamaged classification) and multiclass (undamaged and different damaged cases) classification when considering a dataset with very well separated classes (case A). It also offers good performance in the binary classification between damaged and undamaged scenarios and an even better classification performance in the multiclass tasks, where the classes associated with different damage types are correctly separated. In the analysis related to the second dataset (case B), the EER shows an increase with respect to case A, which is attributable to the miss-classification of the damage classes related to the less severe damages. The outcome for case B demonstrates a highly promising classification ability, showing that, even without any starting information about the structural target system, the model can correctly achieve the classification tasks starting from the simple similarities characterizing these audio signals and their frequency contents. The corruption of the audio dataset with additional external disturbances, like noise and music (MUSAN), did not impact the final accuracy of the model in case A, while in case B, it helps the classification task both in the binary and multiclass cases.

References

1. Farrar, C.R., Worden, K.: Structural Health Monitoring: A Machine Learning Perspective. Wiley, Hoboken (2012)
2. Pan, S.J., Yang, Q.: A survey on transfer learning. IEEE Trans. Knowl. Data Eng. **22**(10), 1345–1359 (2009)
3. Lu, J., Behbood, V., Hao, P., Zuo, H., Xue, S., Zhang, G.: Transfer learning using computational intelligence: a survey. Knowl.-Based Syst. **80**, 14–23 (2015)
4. Bengio, Y.: Deep learning of representations for unsupervised and transfer learning. In: Proceedings of ICML Workshop on Unsupervised and Transfer Learning, pp. 17–36 (2012)
5. Bengio, Y., Bastien, F., Bergeron, A., Boulanger-Lewandowski, N., Breuel, T., Chherawala, Y., Cisse, M., Côté, M., Erhan, D., Eustache, J., et al.: Deep learners benefit more from out-of-distribution examples. In: Proceedings of the Fourteenth International Conference on Artificial Intelligence and Statistics, pp. 164–172 (2011)
6. Caruana, R.: Multitask learning. Mach. Learn. **28**(1), 41–75 (1997)
7. Hinton, G.E., Osindero, S., Teh, Y.-W.: A fast learning algorithm for deep belief nets. Neural Comput. **18**(7), 1527–1554 (2006)
8. Erhan, D., Bengio, Y., Courville, A., Manzagol, P.-A., Vincent, P., Bengio, S.: Why does unsupervised pre-training help deep discriminant learning? J. Mach. Learn. Res. **11**, 625–660 (2009)
9. Lee, H., Pham, P., Largman, Y., Ng, A.: Unsupervised feature learning for audio classification using convolutional deep belief networks. Adv. Neural Inf. Process. Syst. **22**, 1096–1104 (2009)
10. Thomas, S., Seltzer, M.L., Church, K., Hermansky, H.: Deep neural network features and semi-supervised training for low resource speech recognition. In: 2013 IEEE International Conference on Acoustics, Speech and Signal Processing, pp. 6704–6708. IEEE, New York (2013)
11. Veselý, K., Karafiát, M., Grézl, F., Janda, M., Egorova, E.: The language-independent bottleneck features. In: 2012 IEEE Spoken Language Technology Workshop (SLT), pp. 336–341. IEEE, New York (2012)
12. Beigi, H.: Speaker recognition. In: Fundamentals of Speaker Recognition, pp. 543–559. Springer, New York (2011)
13. Balsamo, L., Betti, R., Beigi, H.: A structural health monitoring strategy using cepstral features. J. Sound Vib. **333**(19), 4526–4542 (2014)
14. Chung, J.S., Nagrani, A., Zisserman, A.: Voxceleb2: deep speaker recognition (2018). Preprint. arXiv:1806.05622
15. Johnson, E.A., Lam, H.-F., Katafygiotis, L.S., Beck, J.L.: Phase I IASC-ASCE structural health monitoring benchmark problem using simulated data. J. Eng. Mech. **130**(1), 3–15 (2004)
16. Snyder, D., Garcia-Romero, D., Sell, G., Povey, D., Khudanpur, S.: X-vectors: robust DNN embeddings for speaker recognition. In: 2018 IEEE International Conference on Acoustics, Speech and Signal Processing (ICASSP), pp. 5329–5333. IEEE, New York (2018)
17. Ananthram, A., Saravanakumar, K.K., Huynh, J., Beigi, H.: Multi-modal emotion detection with transfer learning (2020). Preprint. arXiv:2011.07065
18. Povey, D., Ghoshal, A., Boulianne, G., Burget, L., Glembek, O., Goel, N., Hannemann, M., Motlicek, P., Qian, Y., Schwarz, P., et al.: The Kaldi speech recognition toolkit. In: IEEE 2011 Workshop on Automatic Speech Recognition and Understanding. IEEE Signal Processing Society, Piscataway (2011)
19. Ioffe, S.: Probabilistic linear discriminant analysis. In: European Conference on Computer Vision, pp. 531–542. Springer, New York (2006)

Experimental Evaluation of Drive-by Health Monitoring on a Short-Span Bridge Using OMA Techniques

William Locke, Laura Redmond, and Matthias Schmid

Abstract In bridge health monitoring, experts have sought to reduce costs associated with data acquisition systems through the development of an indirect Operational Modal Analysis (OMA) style approach formally known as drive-by health monitoring (DBHM). DBHM is theoretically more cost effective than direct health monitoring strategies, as it can ideally utilize a less dense array of sensors mounted on a single vehicle to capture modal data across a network of bridges. The methodology has received considerable attention the past few decades, with several studies verifying its system identification and damage detection capabilities. However, there exists a noticeable lack of full-scale studies demonstrating the feasibility of the methodology on bridges shorter than 18.28 m (60 ft). Additionally, no studies have been conducted to compare the system identification capabilities of different OMA techniques under the DBHM framework. To address these gaps in research, this study experimentally investigates the feasibility of employing OMA techniques in DBHM to identify the modal properties of a full-scale bridge with spans shorter than 18.28 m (60 ft). Multiple OMA techniques are compared to identify if there exists an approach that offers superior system identification capabilities for data collected from a passenger vehicle traveling at moderate velocities. The results from the experimental study demonstrate that OMA techniques can be leveraged to successfully identify short-span bridge frequencies in the dynamic response of a passenger vehicle even when time variant and nonlinear system properties are present.

Keywords Structural health monitoring · Drive-by health monitoring · Highway bridges · Operational modal analysis · Frequency domain decomposition

1 Introduction

In engineering, the susceptibility of a structural system to damage is a major concern, as sudden and catastrophic failures have been known to occur when damage goes undetected. To prevent sub-optimal performance and/or sudden failures from occurring, experts have developed structural health monitoring (SHM) techniques that utilize networks of specialized sensors to continuously monitor system health over time. In civil engineering, experts have traditionally performed system identification for SHM by taking advantage of natural excitations introduced by ambient forces (e.g. wind and traffic); this output-only approach to system identification is formally known as Operational Modal Analysis (OMA) [1, 2]. OMA is an attractive system identification approach, as tests do not interfere with a structure's normal operations. OMA also has the advantage of identifying modal parameters that are representative of the physical operational behavior of the structure [1]. A problem with utilizing OMA for SHM of civil structures is that the methodology requires dense arrays of sophisticated sensors, integrated into a high performance measurement chain, to reliably capture changes in modal parameters, which can

W. Locke (✉)
Glenn Department of Civil Engineering, Clemson University, Clemson, SC, USA
e-mail: wrlocke@g.clemson.edu

L. Redmond
Glenn Department of Civil Engineering, Clemson University, Clemson, SC, USA

Department of Mechanical Engineering, Clemson University, Clemson, SC, USA
e-mail: lmredmo@clemson.edu

M. Schmid
Department of Automotive Engineering, Clemson University, Greenville, SC, USA
e-mail: schmidm@clemson.edu

K. Grimmelsman (ed.), *Dynamics of Civil Structures, Volume 2*, Conference Proceedings of the Society for Experimental Mechanics Series, https://doi.org/10.1007/978-3-030-77143-0_12

be expensive [1, 3, 4]. Experts have aimed to address these cost related issues through the development of an indirect OMA style approach that utilizes a less dense network of vehicle mounted sensors to capture modal data across not only an entire structure but also a network of structures; this approach to bridge health monitoring is formally known as drive-by health monitoring (DBHM).

The DBHM methodology was first conceptualized in 2004 when Yang et al. identified numerically that it was feasible to use a vehicle as both an excitation source and receiver to capture bridge frequencies [5]. Subsequently, Yang and Lin experimentally verified the feasibility of the methodology when they captured a principle bridge frequency in the dynamic response of a trailer [6]; later studies detected higher order bridge frequencies and mode shapes through the implementation techniques, such as empirical mode decomposition (EMD) and Hilbert transforms [7–10]. Thus, the DBHM methodology has demonstrated the theoretical capability to identify the same modal parameter information as direct SHM procedures. In addition to the research conducted by Yang and his colleagues, other researchers have conducted studies to further improve upon the indirect methodology's system identification capabilities; for detailed information regarding past DBHM studies, please refer to the comprehensive literature reviews in [11, 12].

Within the referenced DBHM literature, there exists a noticeable lack of full-scale field investigations. Furthermore, of the full-scale studies conducted to date, none has focused on bridges shorter than 18.28 m (60 ft) in length [11–17]. For the DBHM methodology to be considered a truly viable SHM strategy for the US bridge network, its feasibility must be demonstrated for bridges less than 18.28 m (60 ft) in length, as they account for approximately 45% of the bridge inventory in the USA and are currently unsuitable candidates for traditional SHM systems due to costs outweighing returns from proactive maintenance [3, 4, 18]. Therefore, the main objective of this study is to experimentally investigate the feasibility of using DBHM to identify the modal parameters of a full-scale bridge with spans shorter than 18.28 m (60 ft). Furthermore, this study aims to demonstrate that existing OMA techniques can be implemented towards the coupled vehicle-bridge problem to help address system identification issues associated with short-span bridges, such as short time history recordings and vehicle frequencies closely matching identified bridge frequencies [19, 20]. Two OMA techniques, peak picking (PP) and frequency domain decomposition (FDD), are employed to identify if either approach provides superior system identification capabilities under the given research framework. Lastly, this study aims to achieve the aforementioned objectives while using data obtained from a standard passenger vehicle. This approach is unique, as all but one experimental study from the literature have been conducted with specialized carts or heavy trucks. The identification of bridge properties from a passenger vehicle is advantageous, as the vehicle-to-bridge mass ratio remains low and helps demonstrate the feasibility of utilizing data collected from traffic to perform population health monitoring on bridges [21, 22].

The remainder of this study is organized as follows: Sect. 2 provides a brief literature review of OMA techniques and their use in the DBHM paradigm; Sect. 3 describes the bridge and vehicle systems used in this study and discusses the instrumentation plan and data acquisition hardware; Sect. 4 provides an overview of the testing procedures; Sect. 5 discusses the specific OMA techniques employed; Sects. 6 and 7 highlight the data processing and OMA analysis results for uncoupled and coupled system identification, respectively; and lastly Sect. 8 summarizes the results from the study and provides insight into improving system identification for shorter span bridges.

2 OMA Literature Review

As mentioned, system identification is performed in OMA by analyzing the dynamic output response of a structure when exposed to unobserved environmental and operational loads. As a consequence of inputs being non-deterministic, fundamental assumptions govern OMA procedures to ensure the consistency of detected modal parameters. The fundamental OMA assumptions are: unknown inputs behave as a Gaussian white noise, meaning they are stochastic, broadband, and smooth; inputs are uncorrelated and spatially distributed equally across the system; and the system is linear and time invariant. Provided the measured response of a structure can be interpreted as the output of a loosely coupled system consisting of an excitation system and the subject structure acting in series, as seen in Fig. 1a, the frequency response function (FRF) of the coupled system ($H_c(\omega)$) can be represented by the equation below [1]:

$$H_c(\omega) = H_e(\omega) \cdot H_s(\omega), \tag{1}$$

where $H_e(\omega)$ and $H_s(\omega)$ are, respectively, the FRFs of the excitation system and the subject structure. The FRF of each subsystem can be represented as a function of the input and output relationships below:

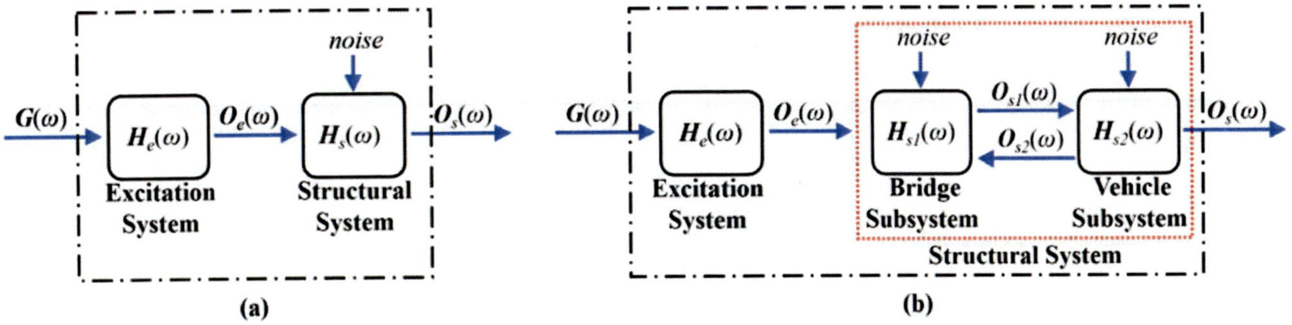

Fig. 1 Example demonstrating the difference between the loosely coupled system model traditionally assumed under for OMA and the strongly coupled system model used in DBHM. (**a**) Conventional loosely coupled system; (**b**) strongly coupled vehicle-bridge system for DBHM

$$H_e(\omega) = \frac{O_e(\omega)}{G(\omega)}, \tag{2}$$

$$H_s(\omega) = \frac{O_s(\omega)}{O_e(\omega)}, \tag{3}$$

where $G(\omega)$, $O_e(\omega)$, and $O_s(\omega)$ represent the Fourier transformations of the assumed Gaussian noise input to the excitation system, the output from the excitation system, and the subject structure's output, respectively [1]. As the output from the structure represents the output response of the coupled system, it contains modal information from both subsystems. Under the traditional SHM paradigm, vehicle-bridge systems closely resemble the framework outlined in Eqs. (1)–(3) and Fig. 1a, where vehicles serve as the excitation system and the bridge response serves as the output of the entire system [1, 23]. Under the DBHM paradigm, however, vehicle-bridge systems create a strongly coupled system where the output from the bridge system feeds back into the vehicle system and the vehicle's response is treated as the entire system's output, as seen in Fig. 1b.

A number of different criteria may apply when classifying OMA techniques, with each criterion highlighting specific benefits and shortcomings common to different techniques in order to guide researchers towards the most appropriate analysis procedures for a given framework. Figure 2 provides a visual breakdown of the classifications for OMA system identification techniques, while information pertaining to the benefits and pitfalls of the classifications can be obtained from [1, 2]. In this study, nonparametric single degree-of-freedom (SDOF) frequency domain techniques are utilized to evaluate the feasibility of employing OMA towards DBHM data. Nonparametric SDOF techniques are selected for this initial study because of their computational efficiency and the presence of a dominant fundamental frequency in most bridge systems [1]; however, future work will include the examination of parametric techniques that can achieve greater predictive accuracy. Section 2.1 provides a brief overview of the OMA techniques examined in this research, while Sect. 2.2 addresses the implementation of the techniques in DBHM.

2.1 Frequency Domain OMA Techniques

The simplest nonparametric OMA technique to implement in the frequency domain is PP. The technique is classified as a SDOF approach due to the assumption that only one mode is present around resonant peaks [2]. PP works by first calculating the power spectral density (PSD) of a system's outputs and then identifying resonant frequencies as the extreme values within the spectrum. The technique can also be leveraged to identify operational deflection shapes and system damping. Despite being simple and intuitive to use, the PP technique is only effective when system damping is low and modes are well separated; however, it is still a useful technique for obtaining initial system identification results in a fast and computationally inexpensive manner [1, 2].

FDD is similar to PP in the sense that modal parameters are estimated from peak frequencies within a subject response spectrum. FDD improves on the PP technique, by leveraging singular value decomposition (SVD) to decompose the cross power spectral density (CPSD) matrix into sets of SDOF systems, thus enabling the detection of closely spaced modes [1, 24]. The detection of closely spaced modes is exact for cases where fundamental OMA assumptions are met, the structure is lightly damped, and the closely spaced mode shapes are geometrically orthogonal; this holds true even under a high level

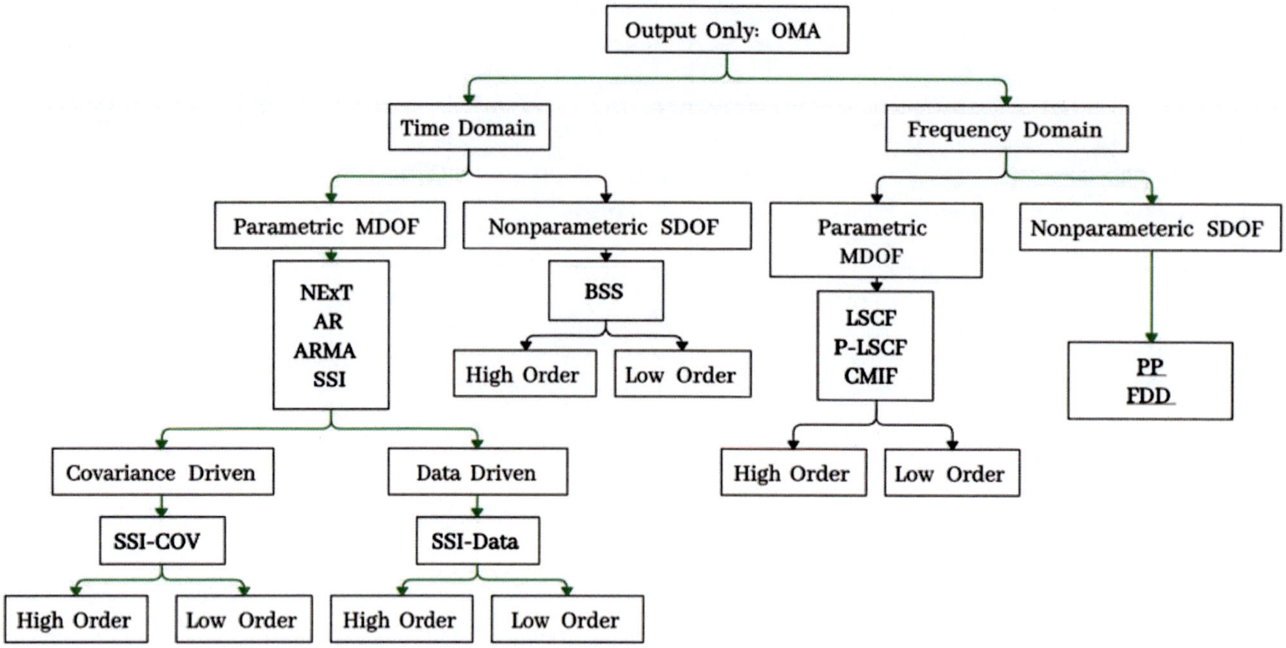

Fig. 2 Hierarchical classification tree for OMA system identification techniques. Underlined techniques are applied to the DBHM data obtained in this study

of noise contamination [24]. For cases in which closely spaced mode shapes are not orthogonal, mode shape estimates become biased against the weaker mode, while estimates for the dominant mode remain close to exact. Even under the condition of biased mode shapes, modal estimates are still considered significantly more accurate than estimates obtained by PP alone [1, 24].

2.2 OMA in DBHM

Despite previous research demonstrating the feasibility of utilizing indirect vehicle measurements to identify modal bridge properties, established OMA system identification techniques, other than PP, have rarely been employed in DBHM due to the fact that vehicle-bridge systems inherently violate the fundamental OMA assumptions [23]. The DBHM methodology violates OMA assumptions in the following ways: (1) vehicle suspension stiffness and damping can be nonlinear; (2) the PSD associated with road surface profiles decreases with increasing frequency, meaning inputs are not Gaussian in nature; (3) loads input at the front and rear wheels are partially correlated; (4) inputs are not spatially distributed equally, as dominant inputs from the road and bridge only act through the wheels; (5) moving vehicle components can introduce strong harmonic excitations: (6) vehicles and bridges have narrowband responses, meaning both systems have identifiable frequencies within the strongly coupled system's output; and (7) the strongly coupled vehicle-bridge system is time variant for the duration of vehicle occupation [25–27].

The aforementioned violations appear to disqualify the application of traditional OMA techniques towards DBHM data at first sight; however, some violations do not significantly affect parameter detection results, while others can be overcome through filtering techniques, sensor placement, and being aware of vehicle properties relative to known bridge properties (e.g. vehicle-to-bridge mass ratios). Violation (1) is not a major concern, as vehicle responses should remain approximately linear in the expected range of motion for a give test speed, meaning models obtained from OMA represent linear approximations of the system at the given working conditions [25]. Violation (2) sometimes results in the masking of bridge frequencies by vehicle frequencies in DBHM, but this can be overcome by carefully controlling vehicle speeds and employing filtering techniques (e.g. EMD) [11]. Violation (3) can be addressed by subtracting the response of multiple identical vehicles simultaneously traveling at the same speed [11, 28]. Another approach to addressing the issue of correlated inputs is to simply analyze the front and rear wheel responses separately. Violation (4) will always be true from the perspective of the bridge system, as the excitation from a vehicle will not uniformly excite the entire bridge or span. From the perspective of the vehicle system, inputs can be approximately spatially distributed equally if the vehicle system of interest is reduced to only the unsprung mass and responses are measured directly on the unsprung mass over each wheel. Violation (5) is partially mitigated by placing sensors on the unsprung mass, as harmonics introduced from the engine and other components are attenuated by the vehicle body and suspension; installing sensors in this manner has also been shown to result in the improved detection of bridge modes [29]. Violation (6) can also be partially addressed by locating sensors on the unsprung vehicle mass, as sprung vehicle frequencies, such as the heave and pitch modes, become masked by more dominate vehicle and bridge frequencies. The issue of narrowband vehicle frequencies interfering with bridge frequency detection can be further overcome by performing OMA testing on the vehicle prior to testing the coupled system [1]. Violation (7) can have a significant impact on OMA results when considering large vehicle-to-bridge mass ratios, as the position of the monitoring vehicle can change the apparent bridge modes; this is especially true for short-span bridges [11, 16, 30]. The violation of the time invariant assumption can be partially mitigated, by controlling the mass ratio and using lighter passenger vehicles when monitoring short-span bridges.

A few studies have been conducted to demonstrate that OMA techniques can be employed in DBHM to successfully identify bridge modal properties [14, 23, 31], but all have focused on longer span bridges in excess of 18.28 m (60 ft). Additionally, no DBHM studies have been conducted to compare the system identification capabilities of multiple OMA techniques. This study aims to address these research gaps by conducting experimental DBHM tests on a 9.14 m (30 ft) long bridge span and analyzing the effectiveness of different frequency domain OMA techniques.

3 Bridge and Vehicle Systems

3.1 Description

The subject bridge in this study crosses over Hard Labor Creek (HLC) along US-221 in Greenwood County, SC. As depicted in Fig. 3, the HLC bridge consists of four simply supported spans, where each span is 9.14 m (30 ft) long and has an edge-to-edge width of 10.21 m (33.5 ft). Exterior span 4 on the south side of the bridge was selected for testing due to the girders being more easily accessible for instrumentation. The primary load bearing system is composed of four reinforced concrete T-Beams installed with a center-to-center distance of 2.43 m (8 ft). The T-Beam construction indicates that the superstructure elements were cast-in-place concurrently with the reinforced concrete deck, meaning the 17.45 cm (6.75 in) beam flanges also act as the deck. The roadway is two lanes wide and has a width of 7.92 m (26 ft). Days prior to testing, the subject section of US 221 was paved with a 7.62 cm (3 in) asphalt overlay; the bridge deck and approach slabs were also freshly paved, allowing vehicles to smoothly enter and exit the bridge without abruptly changing elevation.

During the most recent inspection conducted in March of 2018, inspectors gave the superstructure a condition rating of 6 out of 9 after identifying flexural cracks on the structure. Post dynamic testing, a load rating study conducted on the HLC bridge identified that extensive flexural cracks are present on Exterior span 4 and are continuously propagating due to heavy truck traffic [32]. These findings indicate that crack propagation and breathing may have introduced nonlinear changes in stiffness and damping during dynamic testing that are at least partially responsible for the fluctuations in structural modes observed in Sects. 6.2 and 7.2 below.

The passenger vehicle used in this study is a 2005 RAM 2500 series truck with a quad cab and short bed trim. The truck has a wheelbase of 3.57 m (11.7 ft) and a front and rear track of 1.74 m (5.7 ft). The vehicle's total weight with a driver and passenger is approximately 2722 kg (6000 lbs), with the front and rear weighing 1461 kg (3220 lbs) and 1261 kg (2780 lbs), respectively. The vehicle has an independent front suspension with coil springs and a live axle rear suspension with leaf springs.

3.2 Instrumentation

To be able to monitor the dynamic response of the bridge and vehicle simultaneously, separate data acquisition systems were used. The bridge was instrumented with BDI's STS4 wireless testing system, while the RAM truck was instrumented with a B&K system. Table 1 provides the specifications for the accelerometers and DAQ units of each system. As demonstrated in Fig. 4, a total of three uni-axial accelerometers were installed in two different configurations for the BDI system; this was done to determine if peak frequencies were continuously identifiable under different sensor orientations. Unsprung and sprung sensor configurations were employed for the RAM truck to identify natural vehicle frequencies. For the unsprung

Fig. 3 Aerial and side elevation photographs of HLC bridge. Photographs are provided by the authors of [32]

configuration, sensors were installed on the axles and as close to the center of each wheel as possible; while for the sprung configuration, sensors were installed directly above unsprung sensor locations on the vehicle body. A total of four uni-axial accelerometers were available for testing on the RAM truck, meaning sprung and unsprung tests were conducted separately.

4 Testing Strategy

Experiments for coupled and uncoupled system identification were conducted in this study. Uncoupled system identification experiments were conducted first to identify bridge and vehicle properties prior to DBHM testing. The experiments for uncoupled system identification are listed under items one and two below, while experiments for coupled system identification are listed under item three.

1. *Uncoupled Bridge Testing*: To detect bridge frequencies under true operating conditions without worrying about time constraints and the safety of personnel on the bridge, traffic tests were conducted two days prior to DBHM testing. During testing, the direct bridge response was recorded with ongoing traffic flowing at the posted speed limit of 72.42 kph (45 mph). A total of 20 records were collected for each bridge configuration, with the length of each record being sixty seconds long. With the exception of down time to switch configurations, each record was taken sequentially to ensure minimal variation in bridge properties caused by changing environmental parameters. It should be noted that at the time of testing, Sensor A in the BDI system was malfunctioning; because of this, only the data collected from Sensors B and C were used to obtain the results in Sect. 6.2. This issue was resolved before DBHM testing.

Table 1 Specifications for BDI and B&K data acquisition systems

BDI system					
Accelerometers Type: Model	**Frequency range (Hz)**	**Sensitivity (mV/g)**	**Noise level (μg/Hz)**	**Sample rate [f_s] (Hz)**	**Mounting method**
MEMS: A1521-002	0–400	1000	5	1000	Loctite 410
DAQ Model	**Number of channels**	**Wireless (Y/N)**	**Input range (kHz)**	**Max sample rate (kS/s)**	**Power supply**
STS4-4-IW3	4	Y	Not given	1	Li-Ion battery
B&K system					
Accelerometers Type: Model	**Frequency range (Hz)**	**Sensitivity (mV/g)**	**Noise level (μg/Hz)**	**Sample rate [f_s] (Hz)**	**Mounting method**
Piezo: 4507-B-006	0.2–6000	490 ± 5%	8	1024	Loctite 410
DAQ Model	**Number of channels**	**Wireless (Y/N)**	**Input range (kHz)**	**Max sample rate (kS/s)**	**Power supply**
LAN-XI 3050	6	N	DC-51.2	131	DC to AC power inverter

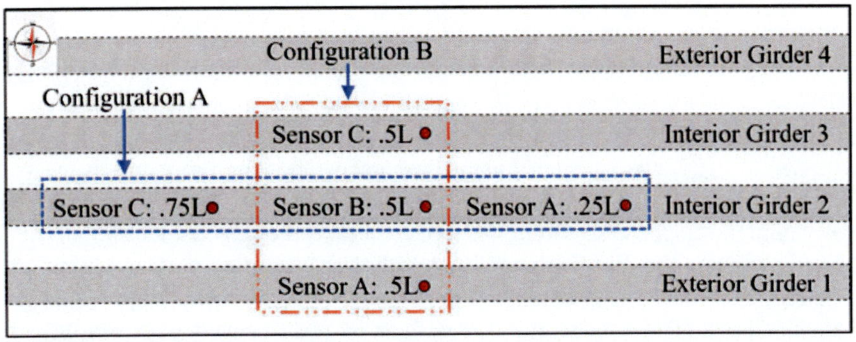

Fig. 4 Configurations for BDI accelerometers

2. *Uncoupled RAM Truck Testing*: To distinguish between vehicle and surface roughness frequencies, road tests were conducted at speeds of 32.19 kph (20 mph), 48.28 kph (30 mph), 64.37 kph (40 mph), and 80.47 kph (50 mph). A total of five records were collected at each speed with sensors installed on the unsprung and then sprung mass (i.e. a total of forty tests were performed); the length of each record was sixty seconds. All tests were conducted on freshly paved sections of US-123 between Clemson, SC and Easley, SC.
3. *Coupled Vehicle-Bridge (DBHM) Testing*: Individual DBHM tests were conducted with only the test vehicle crossing the subject bridge span when it was free of traffic. Testing was conducted concurrently with planned load testing for the SCDOT; therefore, as many tests as possible were performed while the bridge was closed to traffic for this activity. A total of six tests were conducted while vehicle sensors were installed on the unsprung mass and the vehicle was traveling 32.19 kph (20 mph); the length of each record was 30 s. For consistency, the driver and DAQ operator used for RAM road tests were also used for DBHM testing. Direct bridge measurements were also taken during DBHM tests, with the length of each record also being thirty seconds; this data was used to validate the system identification results obtained from the traffic tests conducted days prior. It should be noted that RAM tests were only conducted under Configuration B, meaning Configuration A was not considered during the coupled system analysis. Because the bridge and vehicle DAQ units were not connected, a speed-bump trigger mechanism was employed to help inform both systems when the vehicle was 30.5 m (100 ft) away from the subject bridge span. When the vehicle hit the speed-bump, the vehicle accelerometers recorded the sudden impulse, while a BDI transducer recorded sudden speed-bump displacements and the time of these two events were manually synchronized. It should be noted that excitations from the speed-bump dampened out before the vehicle reached the bridge. Additionally, a person was positioned at the start of exterior span 4 to provide supplemental position data by recording the approximate times the vehicle entered and exited the bridge span.

5 Methodologies

5.1 Peak Picking

In this study, the PSD for bridge and vehicle responses was estimated using the Welch's periodogram method. A moving Hann window with a 66% overlap was employed to enable averaging to reduce spectral leakage within response spectra; the size of the moving window was selected such that a 0.25 Hz spectral resolution was achieved for all data. Resonant frequencies were identified within each response spectra by leveraging Liutkus A.'s PP algorithm based on scale-space theory [33]; a total of ten peaks were selected within each response spectrum. The damping associated with identified resonant frequencies was estimated using the half-power bandwidth method [34]. To identify if the dynamic properties of the bridge and vehicle vary with time, PP was also conducted on the short-time fast Fourier transform (ST-FFT) of response data. The ST-FFT was estimated by employing Matlab's spectrogram function and utilizing the same windowing properties as the Welch periodogram analysis. During this analysis, the ten largest peaks across all windows of the subject spectrogram were identified and their damping was calculated using the half-power bandwidth method.

5.2 Frequency Domain Decomposition

The automated FDD routine developed by Cheynet E. was employed for system identification in this study [35]. Under this methodology, the CPSD needed for SVD was calculated using the same windowing properties as the Welch's periodogram method in Sect. 5.1. Additionally, Liutkus A.'s PP algorithm is pre-coded into the routine to automatically identify the most dominant peak frequencies within the response spectrum; the top ten peaks were selected for all FDD analyses in this study. The referenced methodology calculates the damping associated with each identified peak by fitting an exponential decay to the envelope of impulse response functions obtained via the Natural Excitation Technique (NExT); it can be seen by Fig. 2 that NExT is also classified as an OMA technique [1, 35]. Peak frequencies were also identified from the ST-FDD of response data. The ST-FDD was estimated by dividing time histories into overlapping Hann windows in the same manner as Matlab's spectrogram function. During this analysis, the ten largest peaks across a subject time series were identified and their damping calculated using the NExT procedure as mentioned before.

6 Uncoupled System Identification

6.1 Preprocessing

Preprocessing was performed prior to employing OMA techniques to remove any linear trends, aliasing, and high frequency noise effects. Matlab's digital signal processing toolbox was employed for all signal processing in this study. Initially, detrending was performed to remove linear trends. Then, to decrease the computational cost of OMA procedures, downsampling was employed to reduce sample rates f_s by half the initial values provided in Table 1. After the data was downsampled, aliasing and high frequency noise effects were removed by applying an eighth-order lowpass FIR filter with a passband up to 50 Hz. The 50 Hz cutoff was selected due to an initial analysis of direct bridge data identifying low levels of excitation at higher frequencies. Additionally, DBHM literature indicates that it is difficult to indirectly capture bridge frequencies higher than the second or third mode, which from a preliminary numerical model developed from the bridge plans, were believed to fall below 50 Hz [11].

6.2 Bridge Uncoupled OMA Analysis

The PP and FDD algorithms were used to analyze the averaged and short-time spectrums of the bridge response to ongoing traffic. Figure 5a demonstrates how the PP algorithm identifies obvious resonant frequencies as well as frequencies that have low levels of excitation within an averaged spectrum. To verify identified peak frequencies are associated with bridge modes and not noise, a cumulative sum of the peaks identified by PP in the averaged spectrum was obtained across all records and

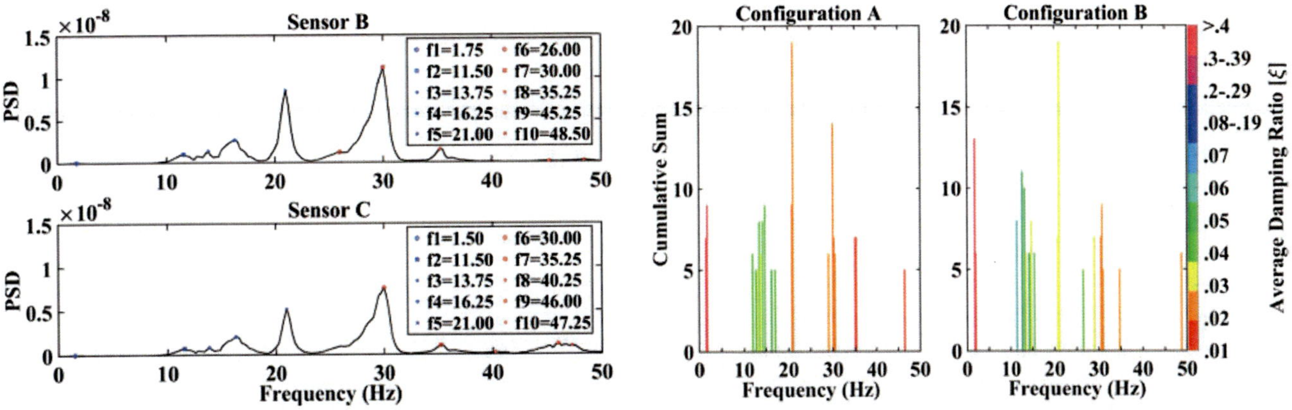

Fig. 5 Uncoupled bridge analysis using PP method across averaged frequency spectra. **(a)** Identified peak frequencies across a single averaged spectrum. **(b)** Cumulative sum of identified peaks across all averaged spectra; plotted in 0.25 Hz wide bins

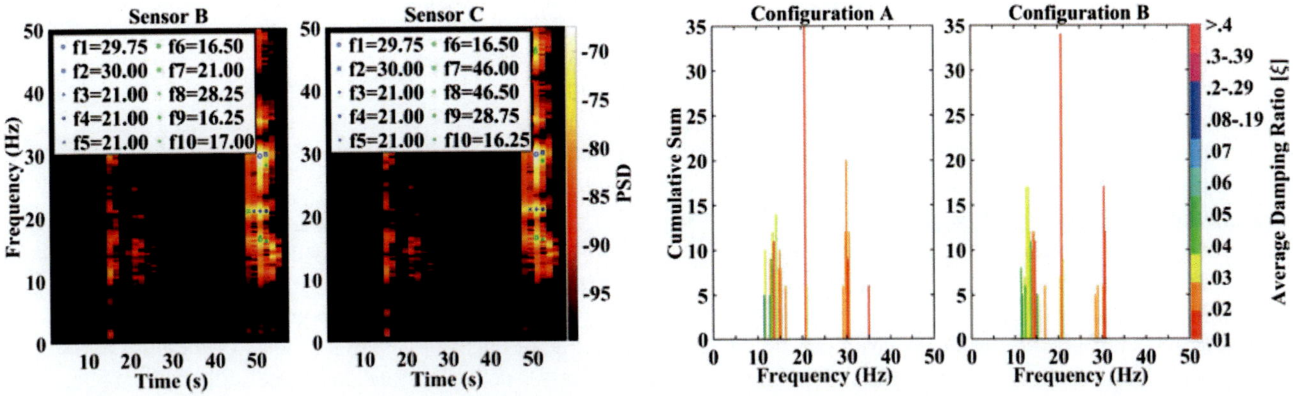

Fig. 6 Uncoupled bridge analysis using PP method across short-time frequency spectrum. **(a)** Identified peak frequencies across a single short-time spectrum. **(b)** Cumulative sum of identified peaks across all short-time spectra; plotted in 0.25 Hz wide bins. *When comparing these results to Fig. 5, it demonstrates that bridge frequencies between 11.5 Hz–16.5 Hz are noticeably influenced by the system's time variant nature, while the bridge frequency at 20.5 Hz–21 Hz is not*

plotted in histograms as shown in Fig. 5b. The width of the histogram bins was set equal to the 0.25 Hz spectral resolution, meaning the cumulative sum in each bin represents the total number of times that specific frequency was identified as a peak. It should be noted that frequencies detected across less than 20% of all tests (i.e. four total times out of the twenty records collected for a given sensor configuration) were considered spurious modes falsely detected by the PP algorithm and were excluded from the plots. Figure 5b demonstrates that resonant frequencies are continuously identified in both configurations around 11.5 Hz–16.5 Hz, 20.5 Hz–21 Hz, 29 Hz–31 Hz, and 34.75 Hz–35.25 Hz, suggesting they are most likely associated with bridge modes; the average calculated damping ratios for these frequencies are, respectively, 4%, 2%, 2%, and 1%, which are in the expected range for a reinforced concrete structure [34]. Additional frequencies are detected around 1.5 Hz–2 Hz, which is in the typical region for vehicle sprung mass frequencies that would have been excited at the posted speed limit [14, 36, 37]. Some additional frequencies are detected at 26 Hz and above 45 Hz; however, the levels of excitation are too low to continuously identify and, therefore, classify.

It is apparent that the bin distributions around 11.5 Hz-16.5 Hz and 29 Hz–31 Hz are wider than at other frequencies, suggesting sources other than noise are responsible for observed short term frequency variations. To verify the observed frequency distributions are the result of system properties varying across test runs and not averaging effects, PP was also performed on the short-time spectrum of bridge response data. Focus was placed on identifying the ten peaks with the highest magnitudes across each short-time spectrum to determine if peaks were continuously detected at the same frequencies for different loading events; again, any frequencies detected less than four total times were excluded. Figure 6a demonstrates how the PP algorithm identifies the highest magnitude frequencies within a single short-time spectrum when multiple loading events are present, while Fig. 6b provides the histograms for total peaks identified across all short-time spectra. As can be seen in Fig. 6b, the frequency bin distributions around 11.5 Hz–16.5 Hz and 29 Hz–31 Hz remain wide, while the distribution around 20.5 Hz–21 Hz remains narrow. These results confirm that the system is time variant, and indicate some

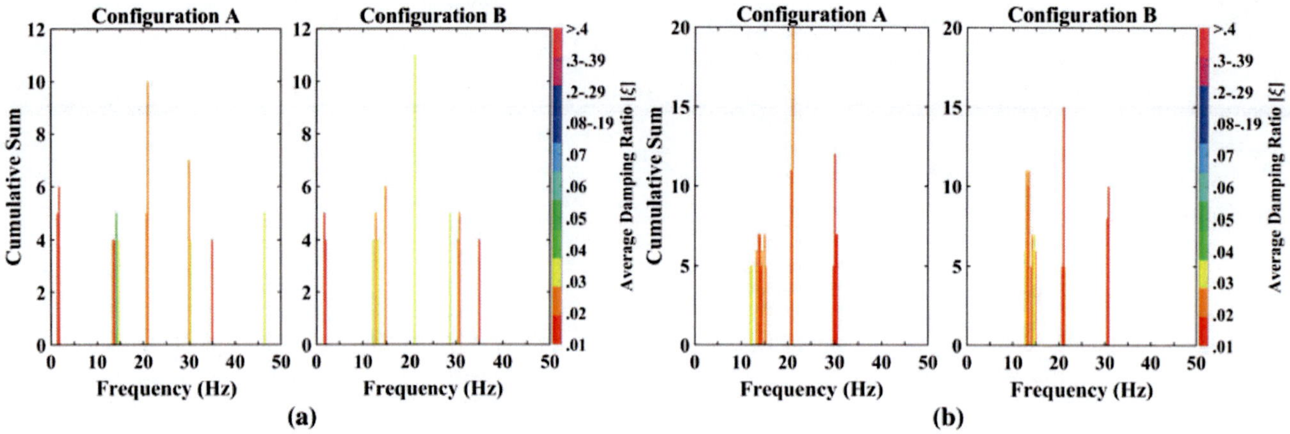

Fig. 7 Cumulative sums from **(a)** FDD and **(b)** ST-FDD analyses of direct bridge data during uncoupled bridge testing; plotted in 0.25Hz wide bins. *Demonstrates narrower frequency bin distributions than base PP method shown in Figs. 5 and 6*

bridge frequencies are noticeably influenced by the system's time variant nature while other frequencies do not appear to be influenced at all.

A partial explanation for the time variant nature of the bridge is that traffic excited breathing in existing cracks, which introduced variations in the bridge's spectral response. Furthermore, tractor-trailers for logging and shipping regularly travel on US-221, meaning trucks similar in size to the HL-93 truck were continuously present during testing and could have caused cracks to completely open. Additionally, the mass of tractor-trailers would have also caused noticeable shifts in certain bridge frequencies, as vehicle-to-bridge mass ratios, assuming HL-93 loads, are estimated to be on the order of 20% [30]. A final explanation for observed frequency variations is that vehicle frequencies were potentially being captured in the bridge response [14]. As vehicles have been shown to introduce higher energy loads in the range of their sprung and unsprung mass frequencies (i.e. 1 Hz–2 Hz and 10 Hz–15 Hz, respectively), it is plausible that these higher energy loads drove the dynamics of the coupled system and caused the bridge to capture vehicle frequencies [14, 36–38]. This explanation is supported by the identification of frequencies at 1.5 Hz–2 Hz and the fact the 11.5 Hz–16.5 Hz distribution is wider than any other observed distribution. It can be noted that the variation in damping in the 11.5 Hz–16.5 Hz region supports all of the above mentioned explanations.

Figure 7 provides the histograms for total peaks identified across all records for the FDD and ST-FDD analyses. To remain consistent with the PP analyses, bin widths were set equal to 0.25 Hz and frequencies detected less than four total times were excluded. As can be seen, the FDD and ST-FDD analyses produce similar results to the averaged and short-time PP analyses; however, the bin distributions between 11.5 Hz–16.5 Hz and 29 Hz–31 Hz in Figs. 5b and 6b are narrower in Fig. 7. The narrower frequency distributions suggests that the FDD technique is more robust against time variant properties when analyzing direct bridge data. The reasoning for this is that by taking the SVD of the CPSD between signals, frequencies continuously detected across all sensors are more pronounced and easily identified as resonant modes, while frequencies introduced by noise or time variant properties not necessarily captured by all sensors are less pronounced and less likely to be identified as peaks. Figure 7 also demonstrates that the damping ratios calculated using the NExT procedure are similar to those calculated using the half-power bandwidth method; however, there is a noticeable difference when considering the frequency bins around 1.5 Hz–2 Hz in Fig. 7a. This difference is believed to be caused by higher mean square errors introduced by initial coefficient estimates when fitting exponential curves to the envelope of impulse response functions for the NExT procedure. During the FDD analyses, average damping estimates were significantly influenced by the choice of initial coefficient estimates, suggesting that the subject NExT procedure is not as robust for identifying damping at certain frequencies and/or excitation levels [1].

6.3 *RAM Truck Uncoupled OMA Analysis*

When analyzing data from the RAM truck road test, emphasis was placed on unsprung data since sensors were only installed on the unsprung masses during DBHM testing. Figure 8 provides an example of a spectrogram plot obtained from road test data collected at 32.19 kph (20 mph). Initially, it can be seen that resonant peaks appear to occur in intervals of approximately

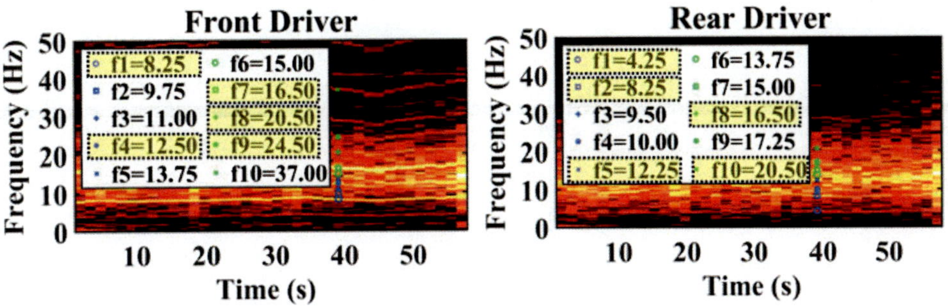

Fig. 8 Spectrogram from 32.19 kph (20 mph) uncoupled RAM test. *Indicates presence of 4 Hz–4.25 Hz frequency intervals*

Table 2 Analysis comparing rotational wheel frequencies to observed spectrogram frequency intervals

Wheel radius: 38.56 cm (15.18 in)			
Velocity [kph (mph)]	Rotational velocity [Hz (RPM)]	Mean observed frequency interval [Hz]	Ratio of observed to rotational
32.19 (20)	3.55 (212.74)	4.00	1.13
48.28 (30)	5.32 (319.12)	6.00	1.13
64.37 (40)	7.09 (425.49)	8.00	1.13
80.47 (50)	8.86 (531.86)	10.00	1.13

3.75 Hz–4.25 Hz; these interval frequencies are believed to be associated with harmonics introduced by wheel defects at the front of the vehicle (e.g. wheel imbalance or radial runout). This conclusion is based on the harmonics being more distinguishable in the sensor data at the front of the vehicle and the PSD of the subject peaks decreasing as the frequencies increase, which is similar to the harmonic behavior associated with wheel nonuniformities [36, 37, 39, 40]. To verify the observed frequency intervals are associated with a defect in one or multiple wheels, Fig. 8 was compared to spectrogram plots obtained from tests conducted at higher speeds. If the interval frequencies are associated with a wheel defect, increases in velocity would have caused wheel RPMs to increase, which in turn would have caused the observed intervals to shift higher into the spectrum and the spacing between peaks to increase. When compared to tests conducted at 48.28 kph (30 mph) and 64.37 kph (40 mph), the identified interval peaks increased in frequency and their spacing increased to approximately 5.75 Hz–6.25 Hz and 7.75 Hz–8.25 Hz, respectively. To further verify these findings, an analysis was conducted where the frequency intervals observed at a given velocity were compared to the rotational speed of the wheels at the same velocity; rotational frequencies were calculated using the nominal unloaded radius of the tires. If the harmonics are associated with wheel speed, the ratio of the observed interval frequencies to the rotational velocity of the wheels will be close to one. Table 2 demonstrates that the subject ratio is approximately 1.13 at all test speeds, indicating that wheel defects at the front of the vehicle are a plausible explanation for the observed harmonic frequencies. The ratios in Table 2 could fall closer to one if the effective radius was employed instead of the unloaded one; however, this property needs to be determined through testing, which was outside the scope of this study. It should be noted that the subject truck is a high mileage all-terrain vehicle used for heavy duty projects, meaning worn elements in the drivetrain or bearings could also be the source of the observed frequency intervals [41, 42]. A more detailed analysis is required to reliably identify the source of the harmonic frequencies; however, this is outside the scope of this project and will be the subject of future model development studies.

As RAM road testing was conducted to identify the frequencies of sprung and unsprung vehicle masses, and because these frequencies are known to fall within small frequency ranges, only the five highest magnitude frequencies were identified during the PP analysis on averaged frequency spectra. During the analysis, histograms for the front and rear of the vehicle were analyzed separately to reduce the issue of correlated inputs introduced by surface roughness. Peak frequencies identified across less than 20% of all tests (i.e. four total times) were again considered spurious and excluded. Additionally, to distinguish between vehicle natural frequencies and the above mentioned harmonics, peak frequencies identified within ± 0.25 Hz of the observed intervals at each respective velocity in Table 2 were also excluded from histogram counts. Excluding frequencies in this manner did allow for vehicle natural frequencies to potentially be excluded as well; this was not a major concern, however, as the natural frequencies would eventually fall outside of the observed intervals as the speed continued to change.

As can be seen in Fig. 9a, the largest peaks are observed at 14 Hz and 13 Hz for the front and rear of the truck, respectively. These frequencies are believed to be associated with the unsprung masses of the truck, as they fall within the 10 Hz–15 Hz frequency range typically associated with passenger vehicles [36, 37]. At the front of the truck, a distribution of frequencies

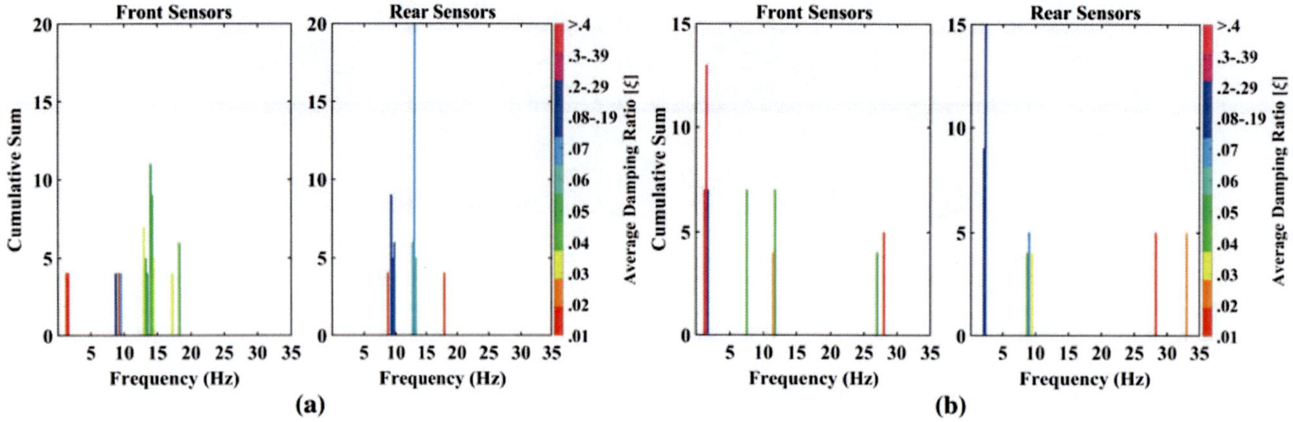

Fig. 9 Analysis of uncoupled RAM truck data using the PP method across averaged frequency spectra. Shows the cumulative sum of identified peaks plotted in 0.25 Hz wide bins for the **(a)** unsprung mass and **(b)** sprung mass

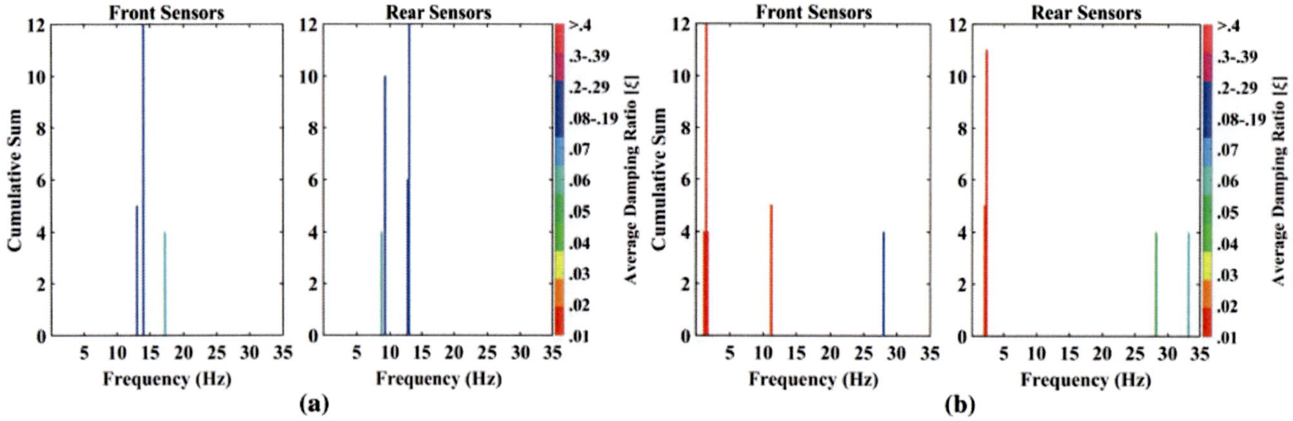

Fig. 10 FDD analysis of averaged frequency spectra during RAM road testing. Cumulative sums are plotted in 0.25 Hz wide bins for **(a)** unsprung mass and **(b)** sprung mass. *Demonstrates narrower vehicle frequency distributions than with PP method shown in Fig. 9*

is also observed around the 14 Hz frequency between 13 Hz–14.75 Hz. The subject frequency distribution can potentially be caused by nonlinear stiffness/damping properties introduced by the supposed wheel defects [36, 37, 39, 40]. Additionally, Fig. 9a demonstrates that a frequency around 1.25 Hz–1.5 Hz is observed at the front of the vehicle; this frequency is believed to be associated with the sprung mass, as it falls within the 1 Hz–2 Hz range for a passenger vehicle's sprung mass frequency and is observed as the most dominate peak for the sprung mass analysis shown in Fig. 9b. Figure 9b further demonstrates that the average calculated damping for the 1.25 Hz–1.5 Hz frequency bins is approximately 20%–29% at the rear of the vehicle, which is also in the range of what is expected for a passenger vehicle suspension [36, 37]. It should be noted that the higher than normal damping ratios at the front of the vehicle in Fig. 9b further supports the conclusion that defects are present at the front of the vehicle. The remaining frequencies in Fig. 9 are believed to associated with vehicle modes (e.g. pitching, rolling, etc.), as other studies have identified similar frequency content when performing EMA and OMA analyses on a truck frame [25]; identifying these frequencies is out of the scope of this project, however, and will be part of the scope of future model development studies.

It should be noted short-time PP and FDD histogram plots were not created for the uncoupled RAM study, as the five largest peaks across each spectrum were almost exclusively associated with the observed harmonic frequencies. Cumulative sums were, however, calculated for a FDD analysis on averaged frequency spectrum data and are displayed in Fig. 10. As can be seen in Fig. 10, the FDD analysis produced similar results to the averaged spectrum PP analysis in Fig. 9, with sprung mass frequencies being identified between 1.25 Hz–1.5 Hz and prominent peaks occurring at 14 Hz and 13 Hz for the front and rear of the vehicle, respectively. Furthermore, in a similar manner to the direct bridge analyses, the FDD analysis appears to be more robust against apparent time variant effects, reducing the frequency bin distribution observed between 13 Hz–14.75 Hz in Fig. 9a to just two bins at 13 Hz and 14.5 Hz in Fig. 10a. Lastly, it can be seen that the NExT procedure also calculates similar damping ratios as the half-power bandwidth method; however, it can again be seen that the methodology struggles to calculate damping in the lower frequency range.

6.4 Summary of Uncoupled OMA System Identification

The results from the uncoupled analysis of the bridge identified that the system violated the time invariant assumption, as nonlinearities believed to be introduced by breathing cracks and heavy tractor-trailers caused the apparent first resonant frequency and damping ratio to fluctuate between 11.5 Hz–16.5 Hz and 1%–5%, respectively. It was observed that the FDD procedures produced narrower variations in the principle frequency, indicating the FDD technique was more robust against apparent time variant system properties. It was also determined, however, that the NExT procedure used for calculating damping under the FDD technique was significantly influenced by initial coefficient estimates and the magnitude of frequency excitations; thus indicating the half-power bandwidth method may be more consistent for identifying damping under the given framework. The results from the uncoupled analysis of the RAM truck also indicated that the vehicle violated the time invariant assumption, as nonlinearities believed to be introduced by nonlinear wheel properties at the front of the vehicle caused notable harmonics and frequency fluctuations in the range of the unsprung natural frequency. In a similar manner to the direct bridge analysis, the FDD technique was robust against the apparent time variant effects, effectively eliminating the frequency bin distribution observed at the front of the truck during the PP analysis. The NExT procedure was once again significantly influenced by initial coefficient estimates and the magnitude of frequency excitations.

7 Coupled Vehicle-Bridge System Identification

7.1 Preprocessing

The same preprocessing procedures used in Sect. 6.1 were used for the coupled analysis when considering direct bridge data. When processing vehicle data, however, time histories were divided into time frames of before the vehicle entered the bridge span and while the vehicle was on the subject bridge span. Figure 11 demonstrates how the full vehicle time histories were divided. On-bridge data encompassed the period of time from half-a-second before the front wheels entered the subject bridge span to half-a-second after the rear wheels exited the span; this correlated to an approximately 2.5 s time window. The half-a-second buffer before and after ensured that the subject time frame fully captured the truck occupation time and allowed multiple Hann windows to be employed for the average and short-time analyses. Before-bridge data encompassed

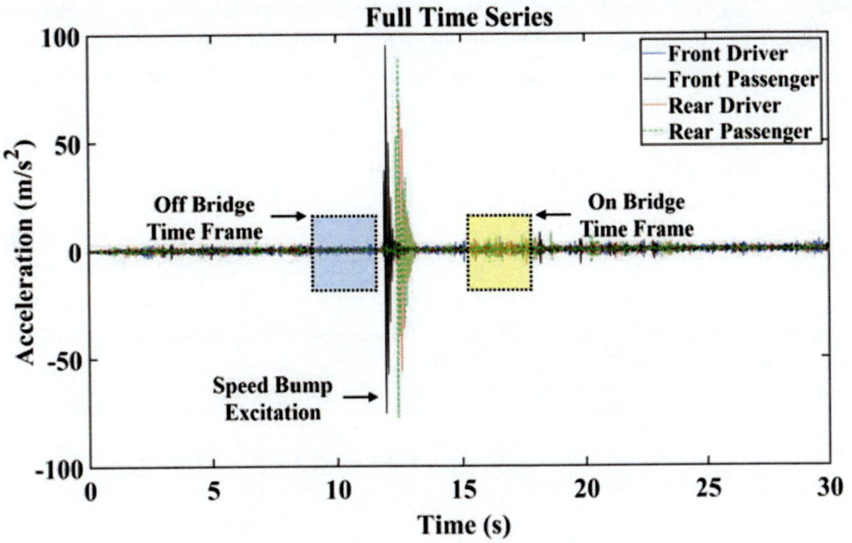

Fig. 11 Example of coupled vehicle time history highlighting before-bridge and on-bridge time frames

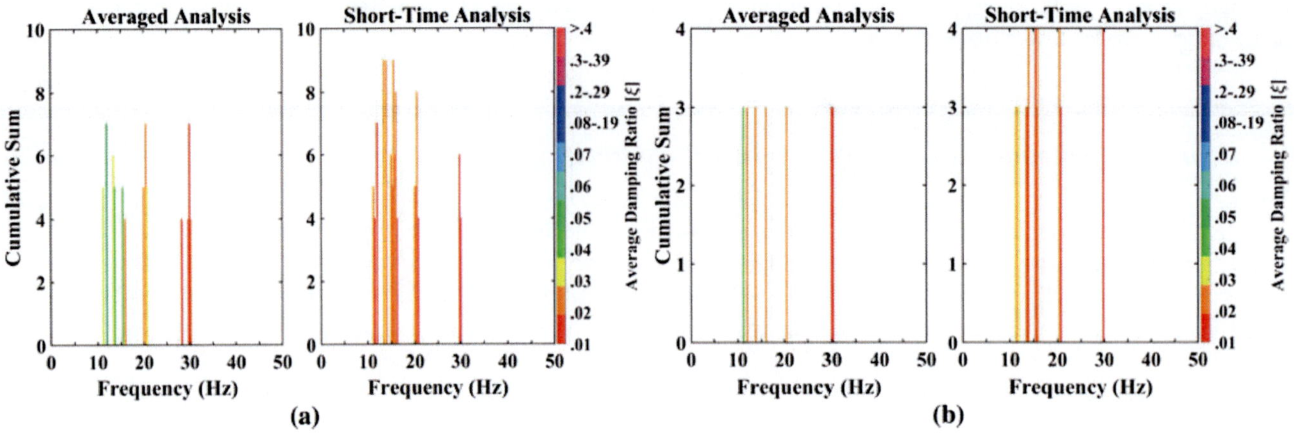

Fig. 12 **(a)** PP and **(b)** FDD analyses of direct bridge data obtained in Configuration B during coupled DBHM testing. Demonstrates smaller groupings of frequency bins in previously observed frequency distribution regions; plotted in 0.25 Hz wide bins

the 2.5 s window of time right before the front wheel hit the speed-bump trigger mechanism; this time frame ensured that the vehicle had already reached the desired test speed and allowed time variant vehicle properties introduced by the initial acceleration to be avoided.

7.2 Bridge Coupled OMA Analysis

As mentioned under item three of Sect. 4, direct bridge measurements were only collected in Configuration B during DBHM testing. Ten peak frequencies were again identified when conducting the averaged and short-time analyses for both the PP and FDD methods; frequencies detected across less than half of all DBHM tests (i.e. less than three total times) were considered spurious and excluded. The use of a larger percentage in this analysis is attributed to the small number of DBHM tests and the fact tests were conducted in a controlled and consistent manner. As can be seen in Fig. 12a, the single vehicle tests appear to excite the same 11.5 Hz–16.5 Hz frequency distribution region previously observed in Figs. 5b and 6b; however, when focusing in on the subject frequency distribution region in Fig. 12a, there appear to be smaller groupings detected between 11 Hz–12 Hz, 13.5 Hz–14.5 Hz, and 15 Hz–16 Hz. The results from the FDD analyses shown in Fig. 12b once again identify narrower distributions, with frequency bins falling around 11 Hz–12 Hz, 13.5 Hz–14 Hz, and 15.25 Hz–16 Hz. Because the vehicle-to-bridge mass ratio was significantly smaller than that of a tractor trailer, it was expected that smaller frequency bin distributions would be observed due to the RAM truck exciting less breathing in flexural cracks. However, the range of the observed groupings is believed to be too wide to have been caused by a single passenger vehicle, suggesting another factor, in addition to vehicle mass and crack damage, introduced frequencies in this region. As mentioned in Sect. 6.2, it is plausible that vehicle frequencies in the region of the principle bridge frequency were captured in the bridge response [14, 38]. The identification of frequencies around the 11 Hz–12 Hz and 13.5 Hz–14.5 Hz regions in Figs. 9 and 10 during the RAM road test analysis provides some validity to this conclusion. An issue with vehicle frequencies potentially being captured in the bridge response is that it becomes difficult to distinguish between the vehicle and bridge; this problem is compounded by the fact the bridge is damaged and, therefore, frequencies are time variant. If vehicle frequencies are being captured in the bridge response, it is believed the issue could be overcome by purposefully utilizing a vehicle configuration with non-interfering resonant frequencies and performing more tests at different speeds to identify which frequencies change and stay the same in the bridge response. This procedure is currently out of the scope of this project, however, as more DBHM testing would be required with the supervision of the SCDOT.

7.3 RAM Truck Coupled OMA Analysis

Due to the small size of the off and on-bridge time frames, and to prevent the removal of frequency bins from being too excessive while also keeping the detection threshold consistent with the analysis in Sect. 7.2, the frequency resolution for

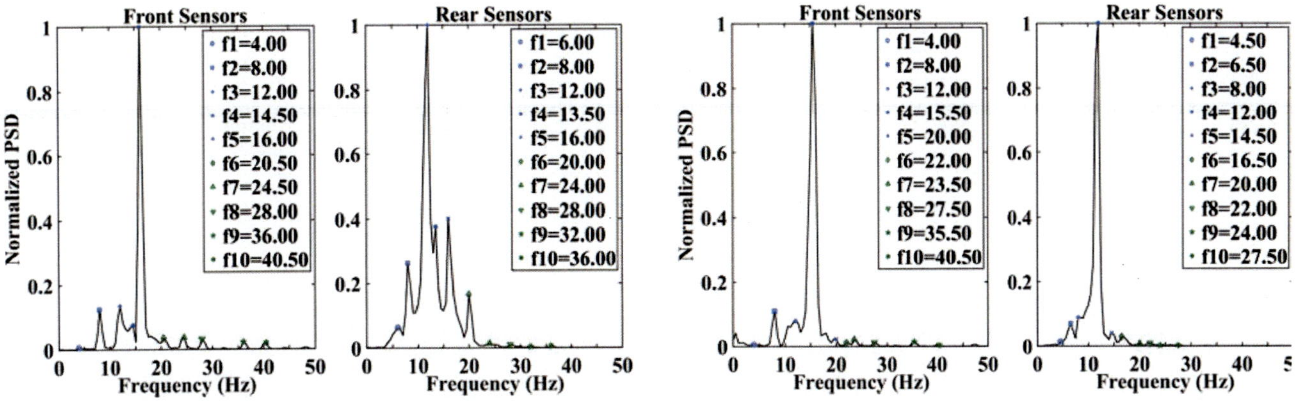

Fig. 13 Averaged FDD analysis of RAM truck data obtained during DBHM testing at 32.19 kph (20 mph). *Demonstrates presence of harmonic frequency intervals and the difference in off and on-bridge spectral content*

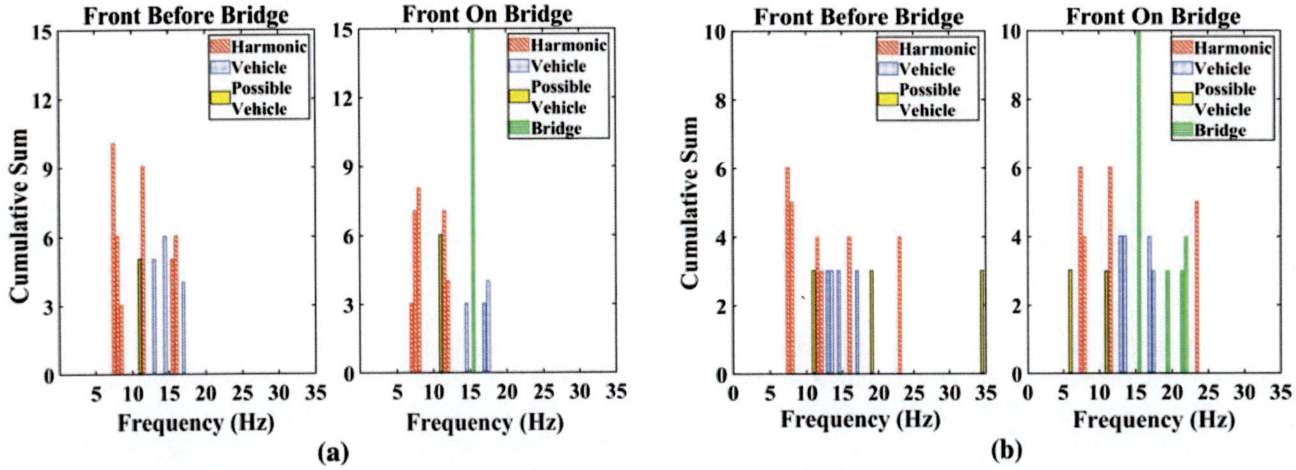

Fig. 14 Short-time analysis of RAM truck DBHM data collected at 32.19 kph (20 mph). Highlights difference between off and on-bridge spectral content using **(a)** PP and **(b)** FDD. *Bins labeled Harmonic, denote likely harmonics from wheel defects. Bins labeled Vehicle, denote frequencies observed on the vehicle during road tests. Bins labeled Possible Vehicle, denote frequencies either observed on the sprung mass or believed to be detected due to increase in identified peaks during analysis. Bins likely associated with bridge frequencies are labeled Bridge*

this analysis was decreased to 0.5 Hz. The width of the histogram bins used for displaying cumulative sums was also set equal to 0.5 Hz to ensure that each bin continued to encompass only a single frequency value. As can be seen in Fig. 13, the harmonic frequencies observed in Sect. 6.3 continue to be detected as dominant frequencies by the PP algorithm. To improve the likelihood of detecting lower amplitude bridge frequencies potentially captured in the vehicle response, the number of peak frequencies identified in the averaged and short-time analyses were increased to ten; this is in-lieu of the five peaks previously used in Sect. 6.3. Because bridge frequencies appear to fall within range of the observed harmonic intervals, the previous approach of excluding the interval peaks to identify natural frequencies of the system could not be conducted. Therefore, an alternative approach was taken where frequencies identified in the range of the observed harmonics or vehicle frequencies were labeled as such, while frequency bins that noticeably changed when the vehicle was on the bridge were labeled as suspected bridge frequencies. Figure 14 provides an example of cumulative count data obtained for the front of the vehicle during the short-time PP and FDD analyses, respectively. It should be noted that damping was excluded from these figures to improve readability. A summary of all notable changes in averaged and short-time spectral content for both the front and rear of the vehicle is provided in Table 3.

Before the front of the truck enters the bridge in Fig. 14, almost all identified peaks can be labeled as harmonic or vehicle frequencies. When the front of the truck transitions from off the bridge to on the bridge in Fig. 14a, the frequency bin at 15.5 Hz effectively triples. Similar results are observed for the FDD analysis in Fig. 14b, where the 15.5 Hz frequency bin does not appear until the front of the vehicle is on the bridge. A change in spectral content is also observed at 19.5 Hz and 21.5 Hz–22 Hz in Fig. 14b and Table 3. The noticeable change in the 15.5 Hz, 19.5 Hz, and 21.5 Hz–22 Hz frequency bins when the front of the truck is on the bridge suggests they are associated with the identified bridge frequencies in Sects. 6.2

Table 3 Summary of notable differences between off and on-bridge spectral content for the RAM truck during DBHM testing

Averaged analysis					
Location (method)	**Subject frequency [Hz]**	**Off bridge sum**	**On bridge sum**	**On to off ratio**	**Mean on bridge damping [%]**
Front (PP)	15.5	3	7	2.33	4.0
Front (PP)	21.5–22	Not observed	3	–	2.5
Rear (PP)	14.5	Not observed	6	–	2.0
Rear (PP)	15.5	Not observed	4	–	4.0
Rear (PP)	19.5	Not observed	3	–	2.0
Front (FDD)	15.5	Not observed	4	–	2.0
Front (FDD)	19.5, 21.5–22	Not observed	3	–	1.0
Rear (FDD)	19.5	Not observed	4	–	1.0
Rear (FDD)	22	Not observed	3	–	1.0
Short-time analysis					
Location (method)	**Subject frequency [Hz]**	**Off bridge sum**	**On bridge sum**	**On to off ratio**	**Mean on bridge damping [%]**
Front (PP)	15.5	5	15	3	2.0
Rear (PP)	14.5	Not observed	7	–	2.0
Rear (PP)	15.5	Not observed	6	–	4.0
Front (FDD)	15.5	Not observed	10	–	2.0
Front (FDD)	19.5, 21.5	Not observed	3	–	2.5
Front (FDD)	22	Not observed	4	–	1.0
Rear (FDD)	14.5	Not observed	4	–	3.0
Rear (FDD)	19.5	Not observed	3	–	3.0

and 7.2. Furthermore, the identified frequency distributions appear to align with the appropriate shifts in bridge frequency that are expected to occur under the DBHM paradigm. Under the DBHM paradigm, the peak frequencies continuously identified around 15 Hz–16 Hz and 20.5 Hz–21 Hz on the bridge are expected to shift in the vehicle response by ±0.5 Hz (3.1 rad) and ±1 Hz (6.1 rad), respectively [43]. Because the frequencies identified from the vehicle are in close agreement with the proper shifts in bridge frequencies, the results indicate that the front of the truck appears to have successfully captured the first two identifiable bridge frequencies. This conclusion is further evidenced by Fig. 13, where both the 15.5 Hz and 22 Hz frequencies appear at the front of the vehicle when it is on the bridge but are not present before the vehicle enters the bridge. An issue with the above observations is that a negative frequency shift is not detected for the lower order 15 Hz–16 Hz bridge frequency but appears to be detected for the higher order 20.5 Hz–21 Hz. The absence of a lower frequency shift can be attributed to a potential resonance matching phenomenon between the bridge frequency and the dominant harmonic frequency observed at 16 Hz. Because the supposed wheel defect introduces a dominant peak at 16 Hz, it is reasonable that a driving force at this frequency would excite resonance in the bridge and allow the bridge excitations to be more dominant in the vehicle response. This explanation is supported by similar observations in DBHM literature and the significant increase in the 15.5 Hz bin counts when the vehicle is on the bridge [5, 11, 38]. Lastly, the damping ratios for the subject frequencies fall between 1%–4% in Table 3, which is in the range of what was observed during direct bridge tests.

Similar changes in spectral content are observed when the rear of the vehicle is on the bridge in Table 3, where bin counts noticeably increase at 14.5 Hz, 15.5 Hz, 19.5 Hz, and 22 Hz. It is plausible the identified frequencies are associated with the bridge, as they fall within ±0.5 Hz (3.1 rad) and ±1 Hz (6.1 rad) of the 15 Hz–16 Hz and 20.5 Hz–21 Hz frequency distributions observed in Sect. 7.2, respectively. Figure 13 provides further support for this conclusion, where frequencies at 14.5 Hz and 22 Hz are detected when the truck is on the bridge but are not present before the vehicle enters the bridge.

When comparing the results obtained by the PP and FDD methodologies, it is apparent that FDD once again produced narrower frequency bin distributions, allowing for changes in spectral content to be more easily detected. Additionally, the ST-FDD method was able to detect changes around the second bridge frequency region, while the ST-FTT method only identified changes around the first bridge frequency; this observation suggests that the ST-FDD approach provides a slight advantage over the ST-FTT method when trying to identify higher order bridge modes captured in the vehicle response. The half-power bandwidth method and the NExT procedure yielded similar results when comparing the average damping ratios for frequency bins below 20 Hz; this is true even at frequencies below 5 Hz where the NExT procedure struggled to identify ratios greater than 1% in Sect. 6. It is also observed, however, that the NExT procedure struggled to calculate damping for frequencies above 20 Hz. The inconsistency of the NExT procedure across Sects. 6 and 7 suggests the half-power bandwidth method might be the better methodology for estimating bridge damping ratios from vehicle response data.

7.4 Summary of Coupled OMA System Identification

The results from the OMA analysis of the direct bridge data identified that the single vehicle excited the same resonate frequencies that were identified during uncoupled testing. When focusing on the previously identified 11.5 Hz–16.5 Hz frequency distribution region, however, smaller frequency groupings were identified rather than a continuous distribution of frequency bins. The identification of smaller groupings instead of a continuous distribution indicated that another factor, in addition to vehicle mass and crack damage, was introducing frequencies in this region. From a review of DBHM literature, it was concluded that vehicle frequencies were potentially being captured in the bridge response. This conclusion was partially supported by the fact that sprung vehicle frequencies were detected in the bridge response during the uncoupled bridge analysis. When analyzing the RAM truck data, a comparison was made between the spectral content of off and on-bridge data to identify if any noticeable changes occurred in the region of suspected bridge frequencies. The results from the OMA analysis of the DBHM data identified that when the front of the RAM truck was on the bridge, noticeable changes in spectral content were observed at 15.5 Hz, 19.5 Hz and 21 Hz–22 Hz. Additionally, when the rear of the vehicle was on the bridge, noticeable changes in spectral content occurred at 14.5 Hz, 15.5 Hz, 19.5 Hz and 22 Hz. It was concluded that the subject frequencies were associated with the 15 Hz–16 Hz and 20.5 Hz–21 Hz frequency distributions observed directly on the bridge, as they all appeared to align with the appropriate shifts in bridge frequency that were expected to occur under the DBHM paradigm. It should be noted that a definitive conclusion could not be made regarding which system the 11 Hz–12 Hz and 13.5 Hz–14 Hz frequencies observed in the direct bridge data belonged to, as frequencies in these regions were continuously detected before the vehicle entered the bridge and while it was on the bridge. Subtle changes in total counts were observed at these frequencies when the vehicle entered the bridge; however, the changes were too inconsistent to conclude if the bridge was responsible.

8 Conclusion

This study is the first to not only successfully demonstrate DBHM on a physical bridge with a span shorter than 18.28 m (60 ft) but also recommend data processing techniques to overcome system identification issues that are more likely to occur when deploying DBHM on short-span bridges. For the 9.14 m (30 ft) long bridge span considered in the experiment, the use of histograms showing the cumulative sum of peak frequencies identified across multiple tests greatly improved the detection of the bridge fundamental frequencies in the presence of closely spaced, time variant, bridge and vehicle frequencies. Additionally, results from leveraged OMA techniques (i.e. PP and FDD) were compared in both averaged and short-time frequency spectra to identify if one approach offered superior system identification capabilities. Both the PP and FDD methodologies were successfully employed to identify what appeared to be the first two bridge frequencies within the dynamic response of the truck, despite the presence of harmonic frequencies and nonlinearities introduced by suspected wheel defects and bridge damage. When comparing results from the PP and FDD analyses, it was determined that the FDD method provided a slight system identification advantage due to it being more robust against apparent time variant properties.

During the experimental and analytical procedures, a number of measures were taken to reduce problems associated with the coupled vehicle-bridge system violating fundamental OMA assumptions. To prevent correlated surface inputs from causing the false identification of natural frequencies, the front and rear acceleration responses of the vehicle were analyzed separately. Additionally, to improve the detection of bridge frequencies while also reducing the presence of harmonics from the engine and other components, sensors were installed on the unsprung masses of the vehicle. The detection of bridge frequencies was further improved by considering the cumulative sum of peak frequencies identified across multiple tests and plotting them in histograms; this approach enabled the detection of lower amplitude and/or closely spaced frequencies that may have gone undetected otherwise.

Several apparent issues were observed in this study that are more likely to be present when deploying DBHM on short-span bridges. Specifically, the observed frequencies were highly effected by the time variant nature of the vehicle and bridge, and the bridge appeared to capture vehicle frequencies within its dynamic response. The latter impeded the identification of the principle bridge frequency. However, these issues did not significantly impact the DBHM results, as obvious changes in the vehicle's spectral content were detected in the regions of suspected bridge frequencies during both the on-bridge PP and

FDD analyses. Considering the cumulative sum of peak frequencies identified across multiple tests was crucial in overcoming these apparent issues. For further improvement, system identification issues (arising from vehicle frequencies being captured in the bridge response) could be addressed experimentally by purposefully shifting known vehicle frequencies across test runs to identify which frequencies change or remain the same in the bridge response. Furthermore, issues related to vehicle defects can be addressed by performing an alignment, balancing, and tuning suspension elements prior to testing. Due to the complex planning process for experimental bridge tests under SCDOT supervision, test runs performed for this paper did not allow for incorporating the above suggestions.

Acknowledgments The authors gratefully acknowledge the support of the National Science Foundation under grant #1633608.

References

1. Rainieri, C., Fabbrocino, G.: Operational Modal Analysis of Civil Engineering Structures, vol. 142, p. 143. Springer, New York (2014)
2. Edwin Reynders. System identification methods for (operational) modal analysis: review and comparison. Arch. Comput. Methods Eng. **19**(1), 51–124 (2012)
3. Lynch, J.P., Loh, K.J.: A summary review of wireless sensors and sensor networks for structural health monitoring. Shock Vib. Dig. **38**(2), 91–130 (2006)
4. Agdas, D., Rice, J.A., Martinez, J.R., Lasa, I.R.: Comparison of visual inspection and structural-health monitoring as bridge condition assessment methods. J. Perform. Constr. Fac. **30**(3), 04015049 (2016)
5. Yang, Y.-B., Lin, C.W., Yau, J.D.: Extracting bridge frequencies from the dynamic response of a passing vehicle. J. Sound Vib. **272**(3–5), 471–493 (2004)
6. Lin, C.W., Yang, Y.B.: Use of a passing vehicle to scan the fundamental bridge frequencies: an experimental verification. Eng. Struct. **27**(13), 1865–1878 (2005)
7. Yang, Y.B., Chang, K.C.: Extraction of bridge frequencies from the dynamic response of a passing vehicle enhanced by the EMD technique. J. Sound Vib. **322**(4–5), 718–739 (2009)
8. Yang, Y.B., Li, Y.C., Chang, K.C.: Using two connected vehicles to measure the frequencies of bridges with rough surface: a theoretical study. Acta Mech. **223**(8), 1851–1861 (2012)
9. Yang, Y.B., Chang, K.C., Li, Y.C.: Filtering techniques for extracting bridge frequencies from a test vehicle moving over the bridge. Eng. Struct. **48**, 353–362 (2013)
10. Yang, Y.B., Li, Y.C., Chang, K.C.: Constructing the mode shapes of a bridge from a passing vehicle: a theoretical study. Smart Struct. Syst. **13**(5), 797–819 (2014)
11. Yang, Y.B., Yang, J.P.: State-of-the-art review on modal identification and damage detection of bridges by moving test vehicles. Int. J. Struct. Stabil.Dynam. **18**(02), 1850025 (2018)
12. Malekjafarian, A., McGetrick, P.J., OBrien, E.J.: A review of indirect bridge monitoring using passing vehicles. Shock Vib. **2015**, **16** (2015)
13. Miyamoto, A., Yabe, A.: Bridge condition assessment based on vibration responses of passenger vehicle. In: Journal of Physics: Conference Series, vol. 305, p. 012103. IOP Publishing, Bristol (2011)
14. Kim, J., Lynch, J.P.: Experimental analysis of vehicle–bridge interaction using a wireless monitoring system and a two-stage system identification technique. Mech. Syst. Signal Process. **28**, 3–19 (2012)
15. Siringoringo, D.M., Fujino, Y.: Estimating bridge fundamental frequency from vibration response of instrumented passing vehicle: analytical and experimental study. Adv. Struct. Eng. **15**(3), 417–433 (2012)
16. Chang, K.C., Kim, C.W., Borjigin, S., et al.: Variability in bridge frequency induced by a parked vehicle. In: Proceedings of the 4th KKCNN Symposium on Civil Engineering, pp. 75–79 (2014)
17. McGetrick, P.J., Kim, C.W.: An indirect bridge inspection method incorporating a wavelet-based damage indicator and pattern recognition. In: Proceedings of the 9th International Conference on Structural Dynamics (EURODYN'14) (2014)
18. Azizinamini, A.: A new era for short-span bridges. Steel Bridge News, pp. 1–2 (2009)
19. Lu, Z.R., Liu, J.K.: Identification of both structural damages in bridge deck and vehicular parameters using measured dynamic responses. Comput. Struct. **89**(13–14), 1397–1405 (2011)
20. Keenahan, J., O'Brien, E.J.: Allowing for a rocking datum in the analysis of drive-by bridge inspections. In: Civil Engineering Research in Ireland, Belfast, 28–29 August, 2014 (2014)
21. Malekjafarian, A.A., Golpayegani, F., Moloney, C., Clarke, S.: A machine learning approach to bridge-damage detection using responses measured on a passing vehicle. Sensors **19**(18), 4035 (2019)
22. Locke, W., Sybrandt, J., Redmond, L., Safro, I., Atamturktur, S.: Using drive-by health monitoring to detect bridge damage considering environmental and operational effects. J. Sound Vib. **468**, 115088 (2020)
23. Yang, Y.B., Chen, W.-F.: Extraction of bridge frequencies from a moving test vehicle by stochastic subspace identification. J. Bridge Eng. **21**(3), 04015053 (2016)
24. Brincker, R., Zhang, L., Andersen, P.: Modal identification of output-only systems using frequency domain decomposition. Smart Mater. Struct. **10**(3), 441 (2001)
25. Swaminathan, B.: Operational Modal Analysis Studies on an Automotive Structure. PhD thesis, University of Cincinnati (2010)
26. Soria, L., Peeters, B., Anthonis, J.: Operational modal analysis and the performance assessment of vehicle suspension systems. Shock Vib. **19**(5), 1099–1113 (2012)

27. Kong, X., Cai, C.S., Kong, B.: Damage detection based on transmissibility of a vehicle and bridge coupled system. J. Eng. Mech. **141**(1), 04014102 (2015)
28. Malekjafarian, A., OBrien, E.J.: Identification of bridge mode shapes using short time frequency domain decomposition of the responses measured in a passing vehicle. Eng. Struct. **81**, 386–397 (2014)
29. Yang, J.P., Chen, B.-H.: Two-mass vehicle model for extracting bridge frequencies. Int. J. Struct. Stabil. Dynam. **18**(04), 1850056 (2018)
30. Kim, C.-Y., Jung, D.-S., Kim, N.-S., Kwon, S.-D., Feng, M.Q.: Effect of vehicle weight on natural frequencies of bridges measured from traffic-induced vibration. Earthquake Eng. Eng. Vib. **2**(1), 109–115 (2003)
31. Yang, Y.-B., Chen, W.-F., Yu, H.-W., Chan, C.S.: Experimental study of a hand-drawn cart for measuring the bridge frequencies. Eng. Struct. **57**, 222–231 (2013)
32. Ziehl, P., Cousins, T., Ross, B., Huynh, N.: Assessment of structural degradation for bridges and culverts. Technical report (2020)
33. Liutkus, A.: Scale-space peak picking (2015). https://hal.inria.fr/hal-01103123
34. Chopra, A.K.: Dynamics of Structures, 5th edn., pp. 174–196. Pearson, New York (2012)
35. Cheynet, E.: Automated frequency domain decomposition (AFDD) (2020). https://zenodo.org/record/4277622
36. Heißing, B., Ersoy, M.: Chassis Handbook: Fundamentals, Driving Dynamics, Components, Mechatronics, Perspectives. Springer Science & Business Media, Wiesbaden (2010)
37. Wang, X.: Vehicle Noise and Vibration Refinement. Elsevier, Amsterdam (2010)
38. Sun, L., Kennedy, T.W.: Spectral analysis and parametric study of stochastic pavement loads. J. Eng. Mech. **128**(3), 318–327 (2002)
39. Walker, J.C., Reeves, N.H.: Uniformity of tires at vehicle operating speeds. Tire Sci. Technol. **2**(3), 163–178 (1974)
40. Pottinger, M.G.: Uniformity: a crucial attribute of tire/wheel assemblies. Tire Sci. Technol. **38**(1), 24–46 (2010)
41. Randall, R.B., Antoni, J.: Rolling element bearing diagnostics—a tutorial. Mech. Syst. Signal Process. **25**(2), 485–520 (2011)
42. Fernandez, A.: Rolling element bearing components and failing frequencies (2020). https://power-mi.com/content/rolling-element-bearing-components-and-failing-frequencies
43. Yang, Y.B., Lin, C.W.: Vehicle–bridge interaction dynamics and potential applications. J. Sound Vib. **284**(1–2), 205–226 (2005)

Investigation of Low-Cost Accelerometer Performance for Vibration Analysis of Bridges

Kirk Grimmelsman

Abstract The use of low-cost MEMS accelerometers for dynamic testing and vibration analysis of civil infrastructure has been steadily increasing since these devices first became commercially available. MEMS accelerometers are available with a variety of different input ranges, sensitivities, and with analog or digital outputs. These accelerometers have become ubiquitous in smartphones, and are also used in the majority of the wireless vibration monitoring systems. MEMS accelerometers with analog output capabilities are of particular interest for long-term SHM applications of civil infrastructure systems since they can be easily integrated with the data acquisition systems employed for the many other types of sensors that are often used for these applications. A low-cost MEMS accelerometer with nominal performance characteristics compatible for most bridge monitoring applications was previously evaluated by the authors to assess its performance and capabilities. The primary focus of the prior study was comparing the frequency characteristics of the MEMS accelerometer against a laboratory-grade accelerometer for specific harmonic excitations provided by a laboratory shaker. The present study is a continuation of the prior evaluation. In this study, several of the low-cost, analog output MEMS accelerometers were installed on a cantilever beam that was subject to free vibrations. The dynamic properties of the beam obtained from the MEMS accelerometer measurements are compared and evaluated against those obtained from more conventional instrument-grade accelerometers also attached to the same structure. The objective of the analysis was to evaluate the performance of the MEMS accelerometers in identifying the dynamic characteristics of the system, including natural frequencies, mode shapes, and damping ratios. Vibration measurements from an in-service highway bridge using both the low-cost MEMS accelerometers and a commercially available MEMS accelerometer were also compared and evaluated. This paper describes the testing program and the data analysis approach and compares the dynamic characterization results from the low-cost MEMS accelerometer and the more expensive instrument-grade accelerometers.

Keywords MEMS accelerometers · Cantilever beam · Highway bridge · Vibration testing

1 Introduction and Scope

A large variety of accelerometers with different sensing technologies, signal conditioning requirements, performance characteristics, and costs are available today from numerous sensor vendors. This can present a challenge in selecting the appropriate accelerometer types in the design of Structural Health Monitoring (SHM) systems for highway bridges. The vibration characteristics are generally known for many short- to medium-span length, multi-beam highway bridges. This class of bridge structure is the most numerous type of bridge in the US highway bridge inventory. Generally speaking, the global structural vibration modes of interest are located in the DC to 50 Hz frequency band and measured vibration amplitudes under operational traffic loads typically less than ±1 g, and rarely ever exceed ±2 g's for this class of structure. There are many types of accelerometers that are capable of meeting or exceeding these general performance characteristics. Micro Electro-Mechanical System (MEMS) accelerometers are increasingly being selected for SHM and vibration testing applications for bridges since they are typically very small and inexpensive, their power requirements are minimal, they can meet the nominal performance requirements for measuring bridge vibrations, and they are available with analog or digital outputs which simplifies integration with various data acquisition design. The primary performance characteristics to

K. Grimmelsman (✉)
Intelligent Infrastructure Systems, Philadelphia, PA, USA
e-mail: kgrimmelsman@iisengineering.com

K. Grimmelsman (ed.), *Dynamics of Civil Structures, Volume 2*, Conference Proceedings of the Society for Experimental Mechanics Series, https://doi.org/10.1007/978-3-030-77143-0_13

consider in selecting a suitable MEMS accelerometer for bridge SHM applications are the measurement range, resolution, noise floor, and type of output.

The characteristics of MEMS accelerometers make them particularly suitable for wireless SHM and vibration monitoring applications for bridges [1, 2]. In long-term SHM applications for highway bridges, a variety of different sensor types such as strain gauges, displacement sensors, tilt sensors, cameras, and weather stations are often included with accelerometers in the instrumentation scheme. The data acquisition designs for such comprehensive SHM systems are most often cabled designs that typically include 24 bit ADC analog input channels which easily accommodate MEMS accelerometers with analog outputs. The available ADC resolution for MEMS accelerometers with digital outputs is commonly less than 24 bits, which is not optimal for measuring low-level vibration signals. An inexpensive MEMS accelerometer was identified by the author to evaluate its suitability for SHM and vibration monitoring applications for short- to medium-span multi-beam highway bridges. This particular MEMS accelerometer is a triaxial sensor with analog outputs so it can be readily integrated into data acquisition systems that can accommodate numerous other types of sensors. This MEMS accelerometer is available for less than $20 and the chip is already mounted on a printed circuit board (breakout board). The sensor can be made field-ready with a minimal electronics expertise and preparation requirements that include basic soldering to add header pins or terminals for the leadwires and mounting the sensor board in a sealed enclosure.

This paper presents the results of a study to investigate the performance characteristics of the abovementioned low-cost MEMS accelerometer. This study is a continuation of a prior effort to evaluate this accelerometer for its suitability for use in SHM and vibration testing applications for highway bridges [3]. In that study, the MEMS accelerometer was evaluated and compared against a conventional instrument-grade accelerometer using harmonic data collected using a shaker and some limited operational vibration data collected from a highway bridge. Time and frequency domain data from these measurements were compared and evaluated. In the present study, measurement data is collected from a small-scale cantilever beam model that is instrumented with the low-cost MEMS sensors and with instrument-grade accelerometers. The primary objective of the current study is to evaluate the modal parameter identification results obtained from vibration measurements collected using the two types of accelerometers. Natural frequencies, damping ratios, and mode shapes are identified for this cantilever beam structure using the low-cost and instrument-grade accelerometers. Operational vibration data collected from a different highway bridge and modal parameter identification results from the low-cost MEMS accelerometer and a different MEMS accelerometer from PCB Piezotronics are compared and evaluated. The test procedures employed, data analysis, and the results obtained from the evaluation program are presented and discussed in the following along with recommendations on the use of the low-cost MEMS accelerometer for highway bridge SHM applications.

2 Low-Cost MEMS and Instrument-Grade Accelerometers

The low-cost MEMS accelerometer evaluated for this study is a sensor that employs the ADXL335 triaxial accelerometer from Analog Devices. The sensor was procured from Adafruit for less than $20 and comes mounted to a printed circuit board and with additional electronic components installed. The only additional preparation required to prepare the sensor for field-use was to solder at screw terminal block onto the header pinouts to facilitate simple attachment of the leadwires, and installing the sensor in a sealed (fully potted) enclosure for actual field installations (Fig. 1). The accelerometer can be powered by up to 5 VDC and has a nominal analog output range of ±3 g. The analog outputs of the accelerometer are ratiometric with the internally regulated 3.3 VDC power. The nominal sensitivity is 330 mV/g for each axis, and *XYZ* filter capacitors give the accelerometer a 50 Hz bandwidth for each direction. The broadband resolution for the accelerometers is 30 micro-g RMS. The noise density for the *X* and *Y* output directions is 150 micro-g/(Hz)$^{0.5}$ RMS and is 300 micro-g/(Hz)$^{0.5}$ for the *Z* output direction. An external 5 VDC power supply was connected to the accelerometer to provide power for the cantilever beam testing whereas the sensor installed on the bridge was powered directly by a 5 VDC output available on the data acquisition hardware. Only one output direction for the accelerometer was evaluated in the study. The *X* output directions was used since the noise density is much better than for the *Z* output direction.

The instrument-grade accelerometer used in conjunction with the low-cost MEMS accelerometer for the cantilever beam testing was the Model 3701G2FA3G (PCB3701) from PCB Piezotronics. This is a variable capacitance accelerometer with a single measurement axis, a sensitivity of 1000 mV/g, a measurement range of +/-3 g, a frequency range of 0–100 Hz (+/-5% accuracy), a broadband resolution of 35 micro-g RMS, and a spectral noise density of 5 micro-g/(Hz)$^{0.5}$ at 100 Hz. The accelerometer can be powered by an excitation voltage of 10–30 VDC.

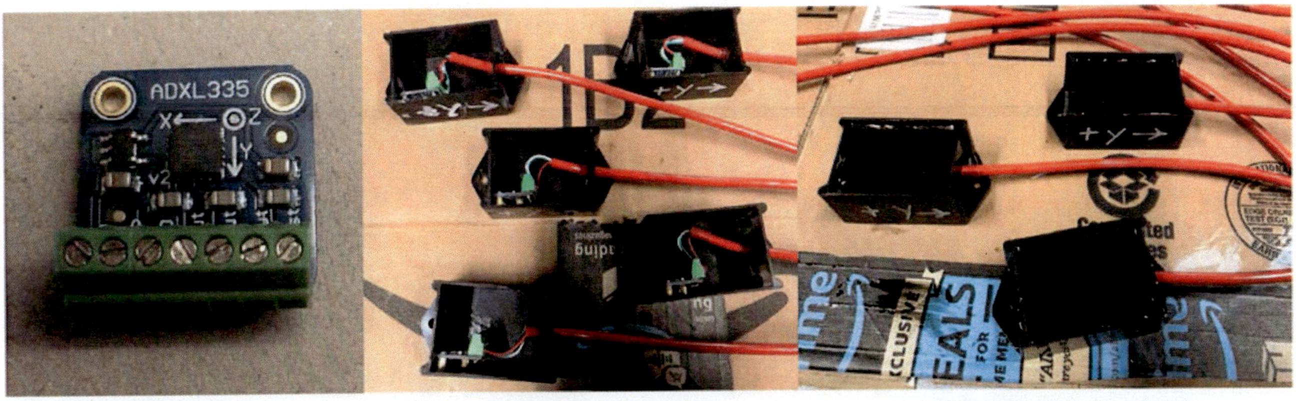

(a) Accelerometer with Screw Terminal Block Installed

(b) Accelerometer with Leadwires in Field Enclosure

(c) Accelerometer Enclosure Sealed in Electronics Epoxy

Fig. 1 Low-cost MEMS accelerometer with screw terminals attached and mounted in field enclosure. (**a**) Accelerometer with screw terminal block installed. (**b**) Accelerometer with leadwires in field enclosure. (**c**) Accelerometer enclosure sealed in electronics epoxy

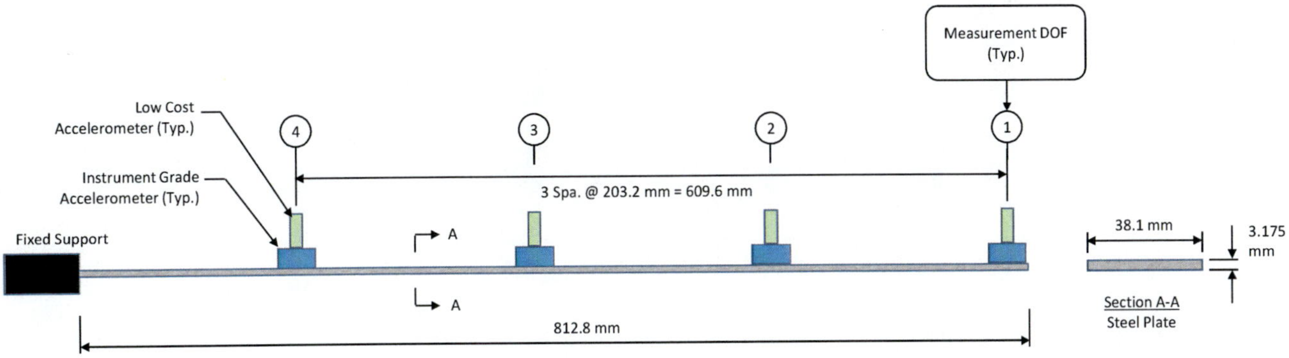

Fig. 2 Schematic of the cantilever beam test setup

3 Cantilever Beam Evaluation

3.1 Test Setup

A small-scale steel cantilever beam structure was developed and instrumented for use in this study. The cantilever beam has a span length of 812.8 mm. A total of four PCB3701 accelerometers were installed with hot glue on the beam at 203.2 mm spaces from the free end. A total of four ADXL335 accelerometers were installed directly on top of the Model 3701 accelerometers with hot glue. A schematic of the cantilever beam setup is shown in Fig. 2 and a photograph of the test setup is shown in Fig. 3. The accelerometers were connected to a NI-9234, four channel input module from National Instruments for data acquisition. External 5 VDC and 24 VDC power supplies were used to provide the excitation voltage for the ADXL335 and PCB3701 accelerometers, respectively. The measurement data were acquired from the acquisition hardware using SigView software.

Since the author only had access to a single NI-9234 input module for testing the cantilever beam, the data was collected in three separate test cases. All of the accelerometers were left in place on the beam throughout the testing to keep the dynamic characteristics of the beam constant during data collection. In Test Case 1, the four instrument-grade accelerometers (PCB3701) were connected to the input module and free vibration data were recorded. In Test Case 2, the four low-cost MEMS accelerometers (ADXL335) were connected to the input module and free vibration data were recorded. In Test Case 3, the two instrument-grade accelerometers and the two low-cost accelerometers located at Positions 1 and 2 as shown in Fig. 2 were connected to the input module and free vibration data were recorded. Free vibration testing consisted of applying an initial displacement to the free end of the beam and releasing. The test was repeated five times for each test case. The sampling rate used for all testing was 400 Hz, and the free vibration response was recorded in each case for 6 s.

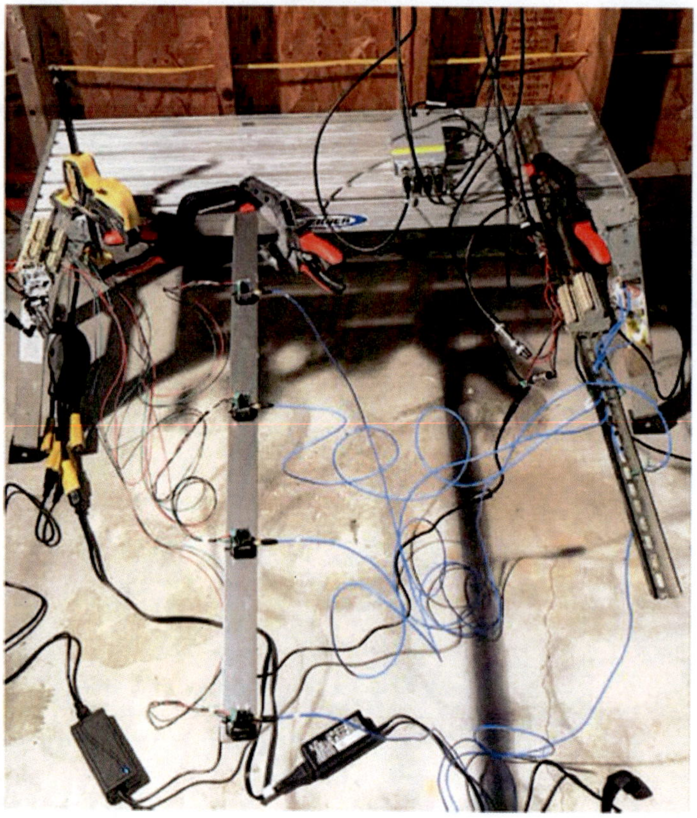

Fig. 3 Photograph of the cantilever beam test setup

3.2 Data Analysis and Results

The vibrations were measured in time domain and FFTs of each output channel were computed in SigView and exported as csv files for analysis in MATLAB. The time domain measurements from the five runs performed for each test case were low-pass-filtered at 100 Hz and then averaged together and plotted for comparison. The FFT spectra from the five runs for each test case were also averaged together and plotted. The averaged time domain and FFT plots for Test Case 1 and Test Case 2 are shown in Fig. 4 and for Test Case 3 in Fig. 5. Modal parameters were identified from the averaged time domain measurements for each test case using a Stochastic Subspace Identification (SSI) function that was coded in MATLAB [4]. The SSI function was configured to identify the first four vibration modes for the beam.

The modal parameters (natural frequencies and damping ratios) identified for each test case are summarized in Table 1. The natural frequency and damping ratio results from Test Case 2 and Test 3 are compared against Test Case 1 as a baseline. Modal Assurance Criterion (MAC) coefficients [5] were computed in MATLAB for the mode shape vectors identified from Test Cases 1 and 2 since these test cases shared the same degrees of freedom. The MAC coefficients obtained for the four vibration modes are as follows: (1) 1.0000 for Mode 1, (2) 0.9997 for Mode 2, (3) 0.9991 for Mode 3, and (4) 0.9988 for Mode 4.

3.3 Discussion

The averaged time domain responses for the three test cases shown in Figs. 4 and 5 appear to be very similar. The difference in the sensitivity characteristic of the PCB3701 (Test Case 1) accelerometers and the ADXL335 (Test Case 2) accelerometers is readily apparent from the measured amplitudes from the sensors that are shown in the figures. The FFT spectra shown in Figs. 4 and 5 are also very similar, except for the larger magnitudes computed for the Test Case 1 measurements. The FFT spectra for Test Case 2 include a spurious peak at approximately 56 Hz. The source of this peak is unknown, and it does not appear in the FFT spectra for Test Case 3 which included measurements from both accelerometer types at measurement

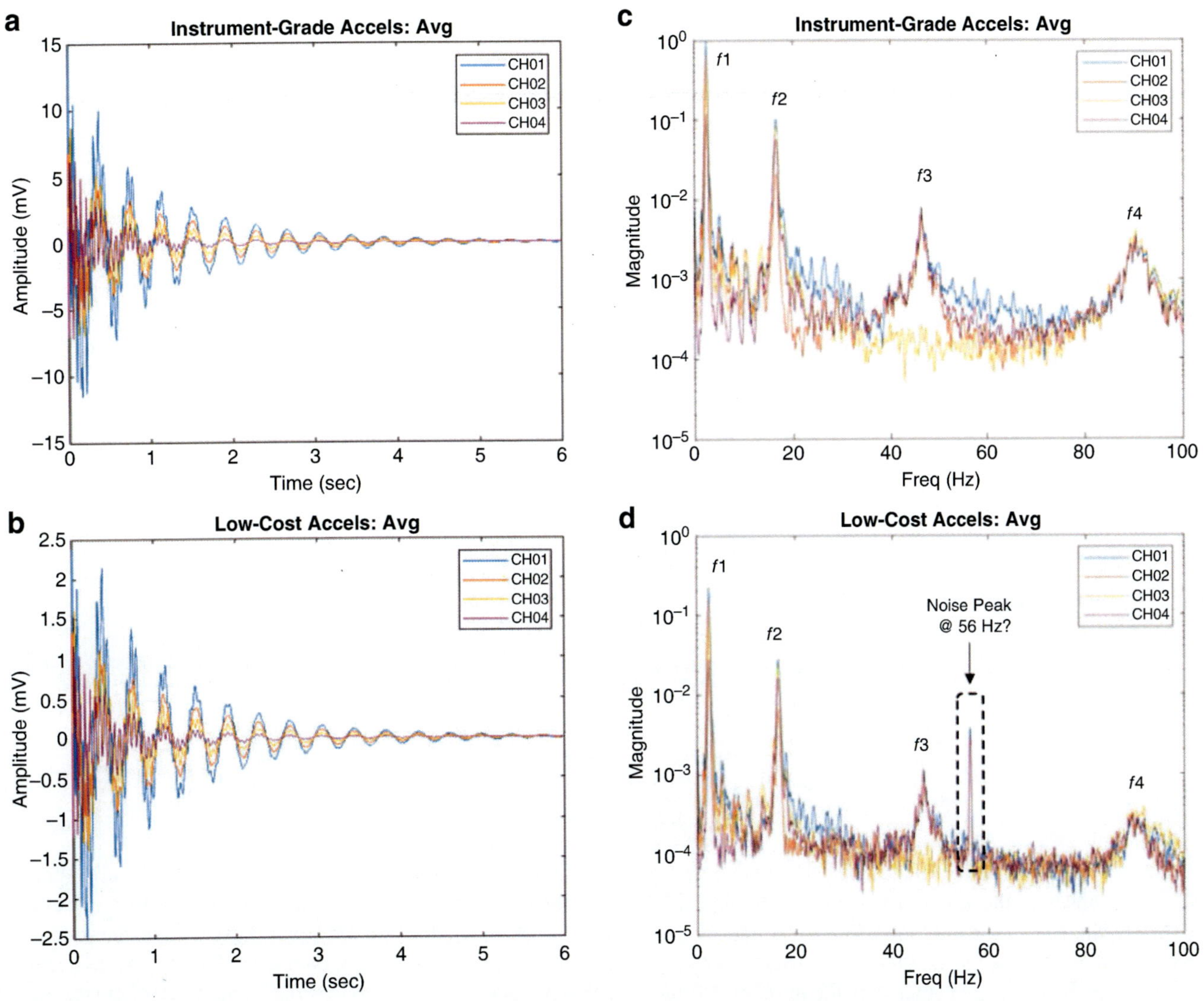

Fig. 4 Average responses: (**a**) time for Test Case 1, (**b**) time for Test Case 2, (**c**) FFT for Test Case 1, and (**d**) FFT for Test Case 2

DOFs 1 and 2 on the cantilever beam. The spike could be the result of electrical noise from the ADXL335 accelerometers at measurement DOF locations 3 or 4. The source of the spike can likely be resolved by conducting further tests with only one ADXL335 accelerometer connected at a time.

The identified modal parameters summarized for the three test cases in Table 1 show very close agreement in the identified natural frequencies for all test cases. The identified natural frequencies from Test Case 2 are all less than 0.1% different than those identified from Test Case 1. Test Case 3, which included both types of accelerometers from measurement DOF locations 1 and 2 on the beam, also shows good agreement with the results of Test Case 1. The differences are all less than 1% even though only two of the four measurement locations on the beam were measured. The damping ratios identified from the three test cases show slightly more variation than what was observed for the natural frequencies, although the differences are generally not so large. The difference in the identified damping ratios between Test Case 1 and Test Case 2 does not exceed 10% for any of the modes. The damping ratios identified for Mode 1 appear to be somewhat high for the steel beam section, but this could be the result of the numerous sensor cables hanging off the beam affecting the free vibration response of its dominant mode. The differences in the damping ratios identified from Test Case 1 and Test Case 3 are also relatively small, with most of percentages being less than 15%. The 35.37% difference in damping ratio for Mode 4 is likely to be due in some part to the inclusion of only two of the four measurement DOF locations on the beam in Test Case 3. It should be noted that the low-cost accelerometer has a filter capacitor on the output that limits the bandwidth to 50 Hz. This did not appear to have any noticeable effect on the capability of these sensors to identify natural frequencies of the beam near 46 and 90 Hz.

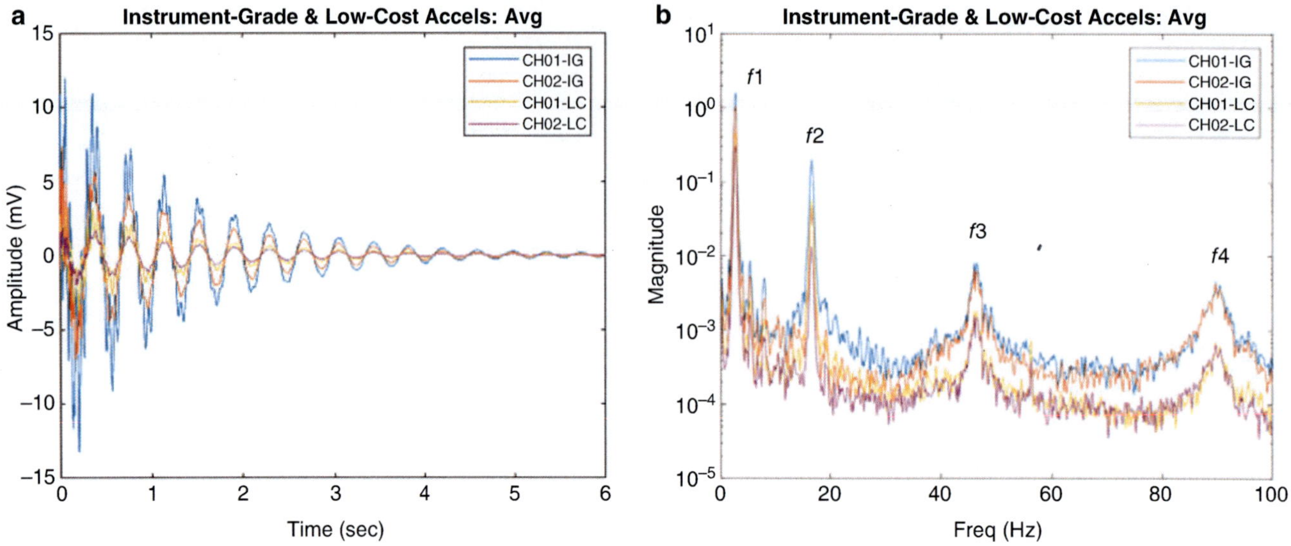

Fig. 5 Average response: (**a**) time for Test Case 3 and (**b**) FFT for Test Case 3

Table 1 Summary modal parameter results from the cantilever beam measurements

Mode	Test Case 1		Test Case 2				Test Case 3			
	f (Hz)	*ζ* (%)	*f* (Hz)	*ζ* (%)	% Diff (*f*)	% Diff (*ζ*)	*f* (Hz)	*ζ* (%)	% Diff (*f*)	% Diff (*ζ*)
1	2.606	4.353	2.606	4.686	0.01	7.66	2.631	3.884	0.95	−10.78
2	16.565	1.821	16.579	1.758	0.08	−3.47	16.574	1.578	0.05	−13.38
3	46.573	1.966	46.587	2.157	0.03	9.75	46.628	1.935	0.12	−1.55
4	90.339	1.346	90.278	1.465	−0.07	8.81	89.472	1.823	−0.96	35.37

The MAC coefficients computed from the mode shape vectors identified from Test Case 1 and Test Case 2 show very good agreement with the smallest value being 0.9988 for Mode 4. Mode shapes were not computed and compared from Test Case 2 since this case only included two of the four measurement DOF locations on the beam. In summary, the modal parameters identified for the cantilever beam show very good agreement when measurement data from the low-cost MEMS accelerometers is compared to that from the instrument-grade accelerometers. There are obvious differences in the sensitivity of the two types of accelerometers, but the performance of the low-cost accelerometer appears to be quite adequate for highway bridge SHM applications based on this evaluation.

4 Highway Bridge Evaluation

A two-span continuous section of a multi-beam highway bridge has been outfitted with a comprehensive SHM system as part of a larger study that is evaluating the effects of truck loads on long-term bridge performance. The sensors deployed on the bridge include both ADXL335 low-cost MEMS accelerometers and a more capable and expensive MEMS DC accelerometer in more limited numbers, in addition to a variety of other sensor types. The highway bridge structure carries two lanes of traffic on I-85 in Troup County, Georgia. The instrumented bridge section has six lines of steel girders which support a cast-in-place composite concrete deck. The length of each instrumented bridge span is approximately 25.2 m. The SHM system includes a total of 15 ADXL335 accelerometers and a total of 5 Model 3741 accelerometers from PCB Piezotronics that are distributed between the two instrumented spans. The Model 3741 accelerometers are considered to be instrument-grade relative to the low-cost ADXL335 accelerometers in this analysis.

The locations of the accelerometers on the instrumented spans are shown in Fig. 6. There is one location in each span where both types of accelerometers are installed at the same location. The SHM system automatically records triggered, short-duration dynamic measurements from the accelerometers and the other sensors whenever a truck crosses over the bridge. The SHM system also records measurement data from the accelerometers on the bridge at 125 Hz continuously for 5 min windows two times each day. The 10 min of continuous vibration measurements recorded from a single day are evaluated in this paper.

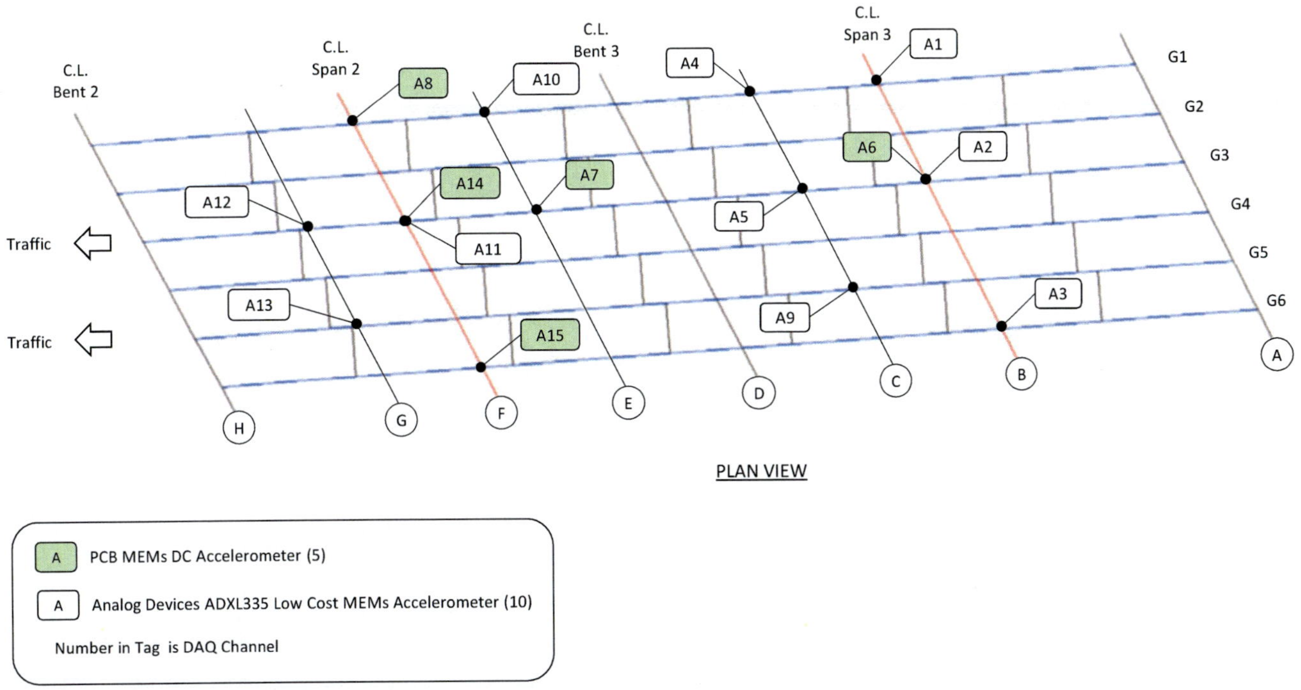

Fig. 6 Accelerometer locations for the SHM system on the Georgia bridge

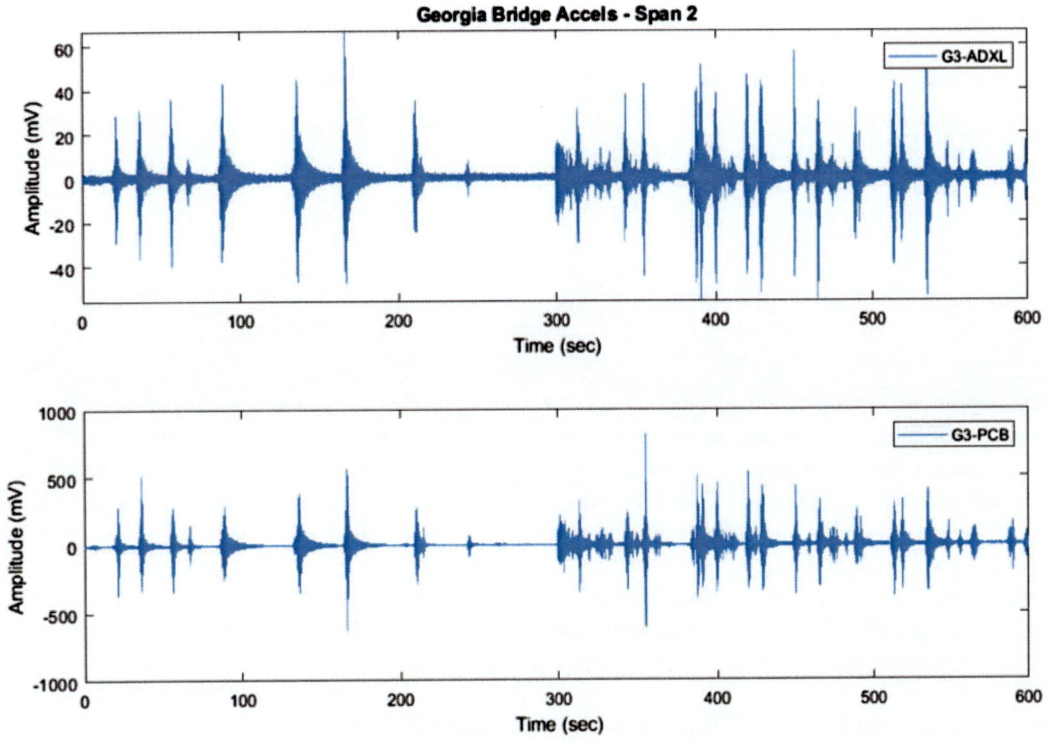

Fig. 7 Typical time domain vibrations measured for the two types of accelerometers that have the same location on Span 2

The typical measured time domain responses for the two types of accelerometers located at the same position on the bridge are shown in Fig. 7 for Span 2 and Fig. 8 for Span 3. The low-cost MEMS accelerometer in each span shows reasonably good agreement with its instrument-grade counterpart in the time domain plots. The difference in the sensitivity between the two types of accelerometers is also readily observable in these two figures.

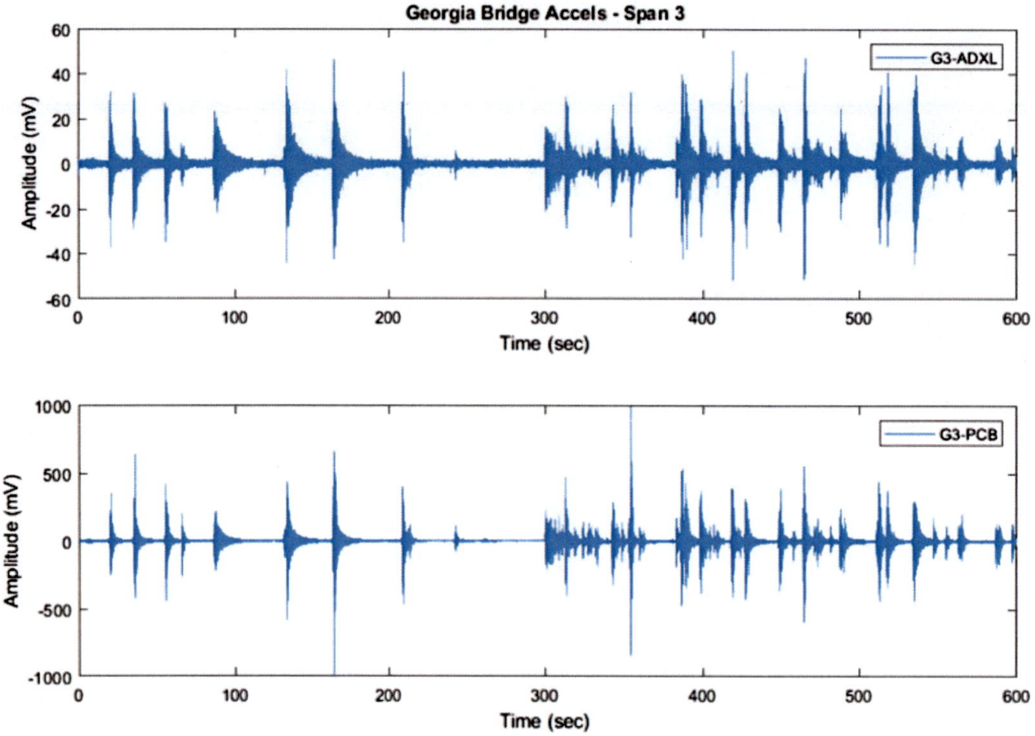

Fig. 8 Typical time domain vibrations measured for the two types of accelerometers that have the same location on Span 3

Table 2 Summary modal parameter results from the accelerometers located in Span 2 of the bridge

Mode	All accelerometers		Instrument-grade accelerometers		Low-cost accelerometers	
	f (Hz)	ζ (%)	f (Hz)	ζ (%)	f (Hz)	ζ (%)
1	3.563	8.8	3.596	12.6	3.486	18.8
2	12.294	5.0	12.711	3.7	13.157	7.5
3	15.404	2.6	16.980	2.4	18.127	49.2
4	19.016	4.9	18.381	1.8	18.788	12.5
5	21.068	2.1	21.384	7.6	19.994	1.9

The vibration measurements from the two types of accelerometers located in Span 2 were analyzed using the same SSI algorithm used with the cantilever beam data to identify natural frequencies and damping ratios. In this case, the SSI function was configured to identify the first five modes from the measurements on Span 2. Results for three different measurement cases were computed: (1) all accelerometers, (2) instrument-grade accelerometers only, and (3) low-cost accelerometers only. Mode shapes were not compared in this case since the two types of accelerometers are located in such different positions on the structure. These identified modal parameter results are summarized in Table 2.

The modal parameters identified from these measurements are generally consistent between the three measurement cases. The natural frequencies identified from the three measurement cases are the most consistent parameter identified. The natural frequency identified for Mode 3 from the low-cost accelerometer only case is the most different from the other two cases. The identified damping ratios show more variation among the three measurement cases, with the value identified for Mode 3 from the low-cost accelerometer only case being most different from the result found for the other two cases. This would appear to indicate that the spatial arrangement of low-cost accelerometers is less than optimal for a reliable identification of Mode 3. Based on the findings from the cantilever beam test, the low-cost accelerometers should be able to provide relatively reasonable modal parameter identification results. This observation is slightly more difficult to confirm from the acceleration measurements on the bridge given their different spatial locations and that the mode shapes for the bridge structure are certainly more complex than those for the simple cantilever beam model. Based on the results obtained from the bridge measurements, there is no clear evidence that the low-cost accelerometers are significantly less capable than the more expensive instrument-grade accelerometers, other than in terms of their smaller sensitivity characteristic. The acceleration measurements acquired from the SHM system on the bridge should really be evaluated and compared in conjunction with modal analysis results obtained from a 3D FEA model of the bridge so that the spatial locations of the various accelerometers can be evaluated for their proximity to nodal points of the mode shapes for each mode. A complete comparison and evaluation of the two types of accelerometers on the bridge requires a set of baseline results with measurement DOFs that are consistent with the locations of both types of accelerometers that are located on the bridge.

5 Conclusions

The performance of a low-cost MEMS accelerometer that appears to be suitable for SHM applications of typical short- to medium-span length, multi-beam highway bridges was evaluated using measurements from a small-scale cantilever beam setup and measurements from a full-scale highway bridge. The measurements recorded by the low-cost accelerometer were compared and evaluated against measurements from more expensive and more capable instrument-grade accelerometers.

The results of the cantilever beam testing showed very good agreement between the modal parameters identified from the two types of accelerometers. Furthermore, these results provide a good amount of confidence that the low-cost accelerometers would be suitable for vibration measurements in a SHM system for a highway bridge. The results obtained from a comparison of the same low-cost accelerometer with a different instrument-grade accelerometer showed reasonably good agreement in time domain, and for identifying natural frequencies. This evaluation was somewhat less conclusive than for the cantilever beam test since the two types of accelerometers were for the most part located at different spatial positions on the structure. The evaluation of the measurements from the highway bridge really needs a consistent baseline of measurement DOF locations to permit a more robust evaluation. The most obvious baseline can be obtained from the analysis of a 3D FE model of the structure. In summary, the performance of the low-cost accelerometers appears to be very promising. However, additional validation studies of these sensors would be beneficial for a comprehensive assessment of their capability for SHM applications for highway bridges.

References

1. Lynch, J.P., Loh, K.J.: A summary review of wireless sensors and sensor networks for structural health monitoring. Shock Vib. Digest. **38**(2), 91–128 (2006)
2. Yu, Y., Ou, J., Li, H.: Design, calibration and application of wireless sensors for global and local monitoring of civil infrastructures. Smart Struct. Syst. **6**(5–6), 641–659 (2010)
3. Grimmelsman, K.A., Zolghadri, N.: Experimental evaluation of low-cost accelerometers for dynamic characterization of bridges. In: Pakzad, S. (ed.) Dynamics of Civil Structures Conference Proceedings of the Society for Experimental Mechanics Series, vol. 2. Springer, Cham (2020). https://doi.org/10.1007/978-3-030-12115-0_19
4. Al-Rumaithi, A.: Subspace identification methods for modal analysis. MATLAB Central File Exchange. https://www.mathworks.com/matlabcentral/fileexchange/69502-subspace-identification-methods-for-modal-analysis (2020). Accessed 5 Nov 2020
5. Allemang, R.J., Brown, D.L.: A correlation coefficient for modal vector analysis. In: Proceedings of the 1st International Modal Analysis Conference, Orlando, pp. 110–116 (1982)

Real-Time Human Cognition of Nearby Vibrations Using Augmented Reality

Elijah Wyckoff, Marlan Ball, and Fernando Moreu

Abstract Human-computer interaction is an important part of structural monitoring and the research community is interested in amplifying human cognition of structural response. Augmenting feedback closes the loop between reality and human, where normally the human would need to check a computer or other separate screen for information on the real-time structural response. The research proposes using an Augmented Reality (AR) application to provide inspectors with real-time information about structural responses, to allow the user to perceive real-time changes. The proposed research seeks to validate the AR application as a useful tool in the field. This paper presents the results of an experiment that show how AR can provide a channel for direct sensor feedback. The application displays a live graph of accelerometer data with a low delay close to real time. The results of the reported experiment prove the usefulness of the AR application in the field by comparing feedback delay over a local area network versus a mobile WiFi hotspot.

Keywords Wireless sensor · Structural response · Vibration · Augmented reality · Real time

1 Introduction

It is important to researchers to monitor vibrations to ensure the safety of the nearby environment. Researchers need to track vibration levels to prevent damage to sensitive machines, but current technology does not allow for a researcher to work freely without having to check a computer monitor or use their hands [1]. Development of machine learning and deep learning algorithms in conjunction with smart sensors enables automated predictive maintenance [2]. This is enabled by IoT technology which is used for wireless sensor networks for environmental sensing and healthcare monitoring [3]. New framework has not been developed for the direct augmentation of data into the user's vision. Researchers have also examined the time delay associated with network communication by analyzing Internet communication features. They propose a control structure to overcome the transmission delay [4]. Fu demonstrates the reliability of a proposed time-delay estimation method through Internet time-delay estimation experiments [5]. This research seeks to investigate the delay in local network and a mobile hotspot for use in the field.

Researchers depend on sensors to inform on critical events especially in structural health monitoring (SHM). Smart Infrastructure wireless sensors are useful for their reliability, low-cost, low power, and fast deployment characteristics [6]. Wireless sensor networks are used for monitoring and assessing vibration risk in historical buildings and cultural sites [7]. Forming a network of wireless sensors supports the gathering of data and decision-making before, during, and after a crisis event. A wireless sensor network in Torre Aquila proved the system is an effective tool for assessing the tower stability while delivering data with loss ratios <0.01% with an estimated lifetime over 1 year [8]. Researchers developed an ad hoc wireless sensor deployment for indoor environmental quality monitoring in office buildings consisting of 19 sensor devices continuously measuring vibration and other levels [9]. Researchers reported installations of wireless sensor networks in a suspension bridge, slab bridge, rail tunnel, and water supply pipeline to manage infrastructure in a way that ensures safe and

E. Wyckoff
Department of Mechanical Engineering, University of New Mexico, Albuquerque, NM, USA
e-mail: ewyckoff13@unm.edu

M. Ball · F. Moreu (✉)
Department of Civil, Construction, and Environmental Engineering, University of New Mexico, Albuquerque, NM, USA
e-mail: mball15@unm.edu; fmoreu@unm.edu

K. Grimmelsman (ed.), *Dynamics of Civil Structures, Volume 2*, Conference Proceedings of the Society for Experimental Mechanics Series, https://doi.org/10.1007/978-3-030-77143-0_14

Fig. 1 AR headset—Microsoft HoloLens

efficient operation by informing data about short- and long-term performance [10]. Wireless sensors are optimal for efficient and reliable data feedback.

Augmented Reality is extremely useful to researchers in informing of real-time data. The concept has been explored in several different areas with the goal of projecting data between the image-based environment and the AR environment. Researchers at the University of New Mexico augmented displacement data, but these values were first recorded and stored in a database before they were graphed in AR [11]. Researchers at Princeton developed a human-machine interface which organizes metadata and provides actionable information by visualizing data about the built environment both on and off-site using AR [12]. Ballor et al. investigated using AR in infrastructure inspections. The framework uses the headset's sensors to capture a high-resolution 3D measurement of the infrastructure which can be used to analyze the state of the structure over time, and track damage progression [13]. Researchers at Los Alamos National Laboratory used AR for structural health monitoring including detecting heat emitted from electronic equipment [14]. AR has a wide range of uses making it a valuable tool for SHM.

The new interface has been explored in the domain of structural design since it is now possible to interface the structural responses with holograms and other models permitting the engineer or technician to quantify structural dynamics in the augmented interface. The interface includes a LEWIS5 (Low-cost Efficient Wireless Intelligent Sensor) which is an Arduino Metro M4 microcontroller equipped with an accelerometer to measure vibrations wirelessly. The sensor requires a power source, which is connected via micro-USB. This data is sent over WiFi using TCP connection to the Microsoft HoloLens AR headset, where acceleration values are plotted real time in the user's field of view. A comprehensive menu is included in the user-interface. This work is innovative in the area of human-structures interfaces, and it enables a new close to real-time control of structures and sensing of dynamics. This research validates the application as a useful tool in the field by comparing the time delay of the graph on a local area network versus a mobile WiFi hotspot.

2 Background

Augmented reality blends interactive digital elements with a real-world environment. An environment is generated by a computer and super-imposed onto the real-world environment, which allows the user to interact with the merged environment. The platform used in this research is optical see-through display, the Microsoft HoloLens. The head-mounted display (HMD) allows for contact-free operation by hand gestures and voice commands. The headset collects data via its image sensors. It also contains an inertial measurement unit, four sensors, an ambient light sensor, a depth camera, a photographic video camera of 2.4-megapixels, and four microphones [15]. The HoloLens has four light-sensitive grayscale cameras labeled in Fig. 1 which track visual features.

The sensing platform is developed to read acceleration data in a triaxial coordinate system as a wireless Structural Health Monitoring (SHM) system. This is done with a Low-cost Efficient Wireless Intelligent Sensor, abbreviated as LEWIS5. The sensor connects via WiFi but requires a power source hooked up via micro-USB. The physical components of the fully assembled sensor are labeled in Fig. 2.

The Metro M4 Express is a 32-bit microcontroller with the ATSAMD51 microchip [16]. The Cortex M4 core runs at 120 MHz with floating point support. The board is powered via micro-USB or barrel jack connection. The Airlift WiFi

Fig. 2 LEWIS5 sensor full assembly; (a) Metro M4 express; (b) MMA8451 accelerometer; (c) Airlift WiFi shield

Shield allows the use of the ESP32 chip as a WiFi co-processor [17]. The Metro M4 microcontroller does not have WiFi built in, so the addition of the shield permits WiFi network connection and data transfer from websites as well as the sending of socket-based commands. The triple-axis accelerometer used for this project is the high-precision MMA8451 with a 14-bit Analog-to-digital converter [18]. Digital accelerometers are used to detect motion, tilt, and basic orientation. For the purpose of this project the accelerometer is used to detect motion as well as vibrations. Its usage range varies from ±2G up to ±8G, which is ideal for its application to this project.

Programming and development of the AR application is done in Unity 2018.4.19f1 taking advantage of the Mixed Reality Toolkit (MRTK) from Microsoft. The application is developed for Universal Windows Platform (UWP) which allows deployment to the HoloLens. The programming platform is Visual Studio 2019, and the code is written in C#. The graph of the live data is developed as a scatter plot, which was chosen as the most effective and efficient solution. The graph is developed based on a tutorial from Catlike Coding [19]. Points at each appropriate coordinate are generated by Unity's default cube game object based on the incoming parsed accelerometer data. Each data point is graphed as a small 3D cube for visual feedback, and the cubes are connected with a Unity LineRenderer so that the graph appears as a line chart. The graph updates with each frame, meaning the cubes are adjusted as time progresses, defined by the function $f(x, t)$. The points are positioned using an Update method with a for loop. A custom matte shader was made for each cube using Unity's material assets. Red designates the x acceleration, yellow designates the y acceleration, and cyan designates the z acceleration.

The application interface consists of six different buttons with specific functionality. From top to bottom: Client Start connects the client to the server via Transmission Control Protocol (TCP). In the context of the application, the computer running the Arduino program acts as the server and the device running the AR application is the client. The Unity code requires the IP Address of the Arduino board, and the Unity code and Arduino code are set up on the same port. Client Stop closes the client connection to the server. The View Start button initiates the Unity function ContinueInput. Incoming data from the server is parsed into x y and z vectors converted to terms of the gravitational constant G, which is graphed on the left side of the interface. The x and y data are also offset so that the x line does overlap and hide the y line. Therefore, the graph axis is labeled as "Z Acc" for the purpose of the experiment as well as simplicity. Included below the graph is a horizontal line indicating the time in seconds of the data. View Stop zeros out the three data lines but does not disconnect the client. The view may be resumed by selecting View Start again. The purpose of the slider at the bottom is a visual aid. When the data breaches the desired threshold value, the line flashes pink as a warning to the user (Fig. 3).

The objective of the experiment was to prove the validity of the real-time acceleration graph and to investigate and verify the minimal delay in the data sending, receiving, and plotting over WiFi on a network and mobile hotspot compared to real time. In the experiment, the researchers used the Microsoft HoloLens 1 AR headset and a Model K2007E01 "SmartShaker," which is an electrodynamic exciter from The Modal Shop. The AR headset was used to augment live acceleration data sent directly from the LEWIS5 sensor. This data was controlled by a set frequency input to the shaker. Therefore, the AR graph could be validated by the known input. Additionally, a video capture allowed for the investigation of the graph delay. This was done with simple frame-by-frame video analysis, which returned an approximate value for the delay in the graph for all two different WiFi connections. The experiment was conducted in a researcher's home office due to restrictions imposed by the global pandemic. The LEWIS5 sensor is positioned horizontal to the ground with the accelerometer z axis matching gravitational acceleration. The sensor is attached to the top of the electrodynamic exciter for controlled input and is connected with a micro-USB cable to a laptop computer as a power source (Fig. 4).

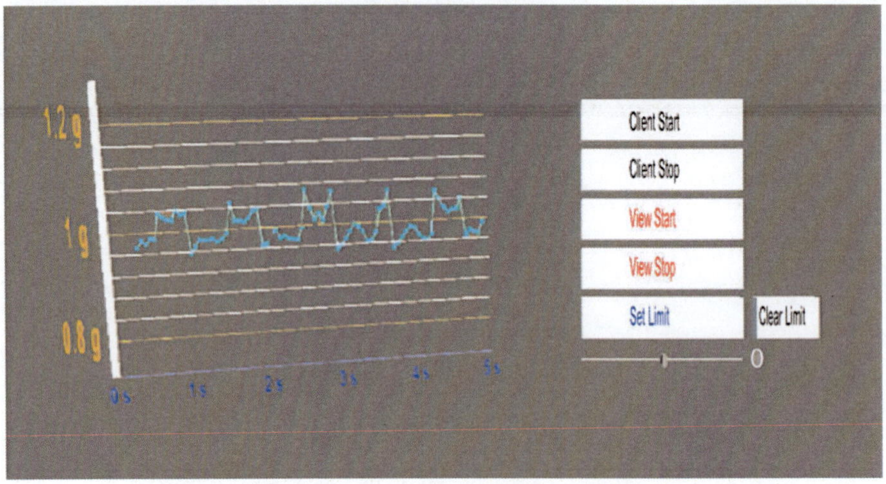

Fig. 3 Application interface as seen in the user's view

Fig. 4 Experimental setup—sensor-shaker configuration

The experiment was conducted following this procedure:

1. Open Arduino IDE and connect LEWIS5 to power source.
2. Attach the LEWIS5 sensor and stinger pairing to the mounting insert on the SmartShaker.
3. Press reset on the sensor once to load the program.
4. Open Serial Monitor and wait for the server to connect to WiFi.
5. Initiate the SmartShaker by selecting a 10 Hz input in a smart phone frequency generator application, which is connected to the shaker via 3.5 mm jack.
6. Once message "Listening for clients on port" is initiated, the Client Start button is selected in the AR application.
7. When the client connects, the Serial Monitor will begin auto scrolling and displaying the three columns of acceleration data, at which time View Start is selected.

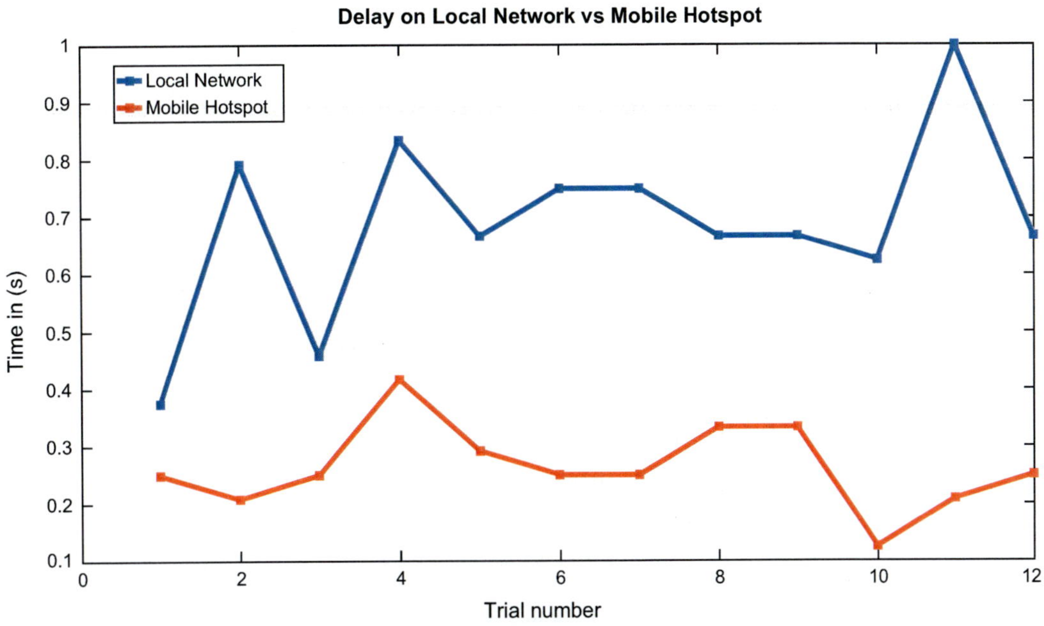

Fig. 5 Delay on local network vs. mobile hotspot

8. The three colored lines are plotted on the graph in the user's view.
9. View Stop and Client Stop are selected to close the graph and shut down the client's connection to the server.
10. The visual feedback from the graph is recorded in the HoloLens as a video to validate the results and analyze the delay.
11. The time-delay data and Arduino data are plotted in MATLAB.

The time delay in the program was investigated using video analysis. By counting the frames between the initial sensor acceleration and the recorded response, the time delay can be approximated. With the known value of the video framerate in FPS, the time delay of the application can be calculated using the following formula.

$$(\text{Frame}_1 - \text{Frame}_0) * \frac{1}{\text{Video FPS}} = \text{Time Delay} \tag{1}$$

3 Test Results and Analyses

The test was run with a total of three trials with four excitations for each WiFi connection. Between each trial the sever-client connection was shut down and the AR application was restarted. Therefore, the graph displayed a total of 12 positive peaks which could be analyzed to determine the time delay. Frame_0 was designated as the time of initial acceleration upwards by the sensor. Thus, Frame_1 designates the response plotted on the graph in the user's view. The difference in frames was recorded and the time delay was calculated with Eq. (1). The results were plotted in MATLAB for comparison in Fig. 5 and the individual values are reported in Table 1.

The results of the experiment validate the AR application for use in the field. As proven clearly in the graph, the time delay on the local network was inconsistent and slow. The time delay in the graph response on the mobile WiFi network is consistently under 0.5 s, proving the reliability of the application for work in the field. The average time delay in the graph on the mobile network is much closer to real time at 0.263 s versus 0.6875 s on the local network. Additionally, the delay on the local network peaks at one full second while the delay peaks at less than 0.5 s on the mobile network.

Table 1 Experimental results

Trial	LAN delay in (s)	Mobile delay in (s)
1	0.3750	0.2500
2	0.7917	0.2083
3	0.4583	0.2500
4	0.8333	0.4167
5	0.6667	0.2917
6	0.7500	0.2500
7	0.7500	0.2500
8	0.6667	0.3333
9	0.6667	0.3333
10	0.6250	0.2500
11	1	0.2083
12	0.6667	0.2500
Average	0.6875	0.2630

4 Conclusions

This paper developed and tested an augmented reality application for live sensor feedback. An analysis of the delay in the acceleration graph shown in AR was conducted and the graph was verified by a known vibration frequency input to the sensor. Researchers demonstrate the reliability of the AR application in graphing acceleration data close to real time. This emphasizes the importance of accurate measurements and transmission consistency. The variance in time delay on the local network makes for a less reliable mode of data transmission in the AR application, whereas the connection on the mobile hotspot relays and graphs vibration data close to real time. The research helps to understand the reliability of several WiFi network options in close to real-time acceleration data transmission while developing an AR tool for data feedback and perception for use in the field.

References

1. Karlsson, M., Hörnqvist, F.: Robot condition monitoring and production simulation. Ph.D. dissertation, Dept. Civilingenjör, Maskinteknik, Luleå Tekniska Univ., Institutionen för teknikvetenskap och matematik, Luleå, Sweden (2018)
2. Namuduri, S., Narayanan, B.N., Davuluru, V.S., Burton, L., Bhansali, S.: Review—Deep learning methods for sensor based predictive maintenance and future perspectives for electrochemical sensors. J. Electrochem. Soc. **167**(3), 037552 (2020). https://doi.org/10.1149/1945-7111/ab67a8
3. Raj, A., Steingart, D.: Review—Power sources for the Internet of Things. J. Electrochem. Soc. **165**(8) (2018). https://doi.org/10.1149/2.0181808jes
4. Yang, S.H., Chen, X., Tan, L.S., Yang, L.: Time delay and data loss compensation for Internet-based process control systems. Trans. Inst. Meas. Control. **27**(2), 103–118 (2005). https://doi.org/10.1191/0142331205tm133oa
5. Fu, X., Fu, X.: A predictive algorithm for time delay Internet network. In: 2011 International Conference on Electronics, Communications and Control (ICECC), Ningbo, pp. 666–669 (2011). https://doi.org/10.1109/ICECC.2011.6066402
6. Morimoto, R.: A socio-economic analysis of Smart Infrastructure sensor technology. Transp. Res. Part C Emerg. Technol. **31**, 18–29 (2013). https://doi.org/10.1016/j.trc.2013.02.015
7. Abruzzese, D., Angelaccio, M., Giuliano, R., Miccoli, L., Vari, A.: Monitoring and vibration risk assessment in cultural heritage via Wireless Sensors Network. In: 2009 2nd Conference on Human System Interactions (2009). https://doi.org/10.1109/hsi.2009.5091040
8. Ceriotti, M., et al.: Monitoring heritage buildings with wireless sensor networks: the Torre Aquila deployment. In: 2009 International Conference on Information Processing in Sensor Networks, San Francisco, CA, pp. 277–288 (2009)
9. Brunelli, D., Minakov, I., Passerone, R., Rossi, M.: POVOMON: an Ad-hoc Wireless Sensor Network for indoor environmental monitoring. In: 2014 IEEE Workshop on Environmental, Energy, and Structural Monitoring Systems Proceedings, Naples, pp. 1–6 (2014). https://doi.org/10.1109/EESMS.2014.6923287
10. Bennett, P., Stoianov, I., Fidler, P., Maksimovic, C., Middleton, C., Graham, N., Soga, K., Hoult, N.: Wireless sensor networks: creating 'smart infrastructure'. Inst. Civil Eng. **162**, 136–143 (2009)
11. Aguero, M., Maharjan, D., Rodriguez, M.D., Mascarenas, D.D., Moreu, F.: Design and implementation of a connection between augmented reality and sensors. Robotics. **9**(1), 3 (2020). https://doi.org/10.3390/robotics9010003
12. Napolitano, R., Liu, Z., Sun, C., Glisic, B.: Combination of image-based documentation and augmented reality for structural health monitoring and building pathology. Front. Built Environ. **5** (2019). https://doi.org/10.3389/fbuil.2019.00050
13. Ballor, J.P., et al.: Augmented reality for next generation infrastructure inspections. In: Barthorpe, R. (ed.) Model Validation and Uncertainty Quantification Conference Proceedings of the Society for Experimental Mechanics Series, vol. 3. Springer, Cham (2019). https://doi.org/10.1007/978-3-319-74793-4_23

14. Morales Garcia, J.E., Gertsen, H.J., Liao, A.S.N., Mascarenas, D.D.L.: Augmented reality for smart infrastructure inspection. Technical report, Los Alamos National Lab, (LANL), Los Alamos, NM, USA (2017)
15. Holmdahl, T.B.: A closer look at the Microsoft HoloLens hardware. Microsoft Devices Blog (2015)
16. Industries, A.: Adafruit Metro M4 feat. Microchip ATSAMD51. https://www.adafruit.com/product/3382 (n.d.)
17. Industries, A.: Adafruit AirLift Shield - ESP32 WiFi Co-Processor. https://www.adafruit.com/product/4285 (n.d.)
18. Industries, A.: Adafruit triple-axis accelerometer - ±2/4/8g @ 14-bit - MMA8451. https://www.adafruit.com/product/2019 (n.d.)
19. Flick, J.: Building a graph. https://catlikecoding.com/unity/tutorials/basics/building-a-graph/ (2020)

Elijah Wyckoff is a graduate research assistant in the Department of Mechanical Engineering at the University of New Mexico, pursuing a MS. His BS is from Arizona State University in the Department of Mechanical Engineering (2020).

Marlan Ball is a research assistant in the Department of Civil, Construction and Environmental Engineering at the University of New Mexico. His BS is from LeTourneau University in Electrical Engineering Technology.

Fernando Moreu is an assistant professor in the Department of Civil, Construction and Environmental Engineering of the University of New Mexico, with courtesy appointments in the Department of Mechanical Engineering and the Department of Electrical and Computing Engineering. His PhD is from the University of Illinois at Urbana-Champaign in Structural Engineering (2015).

Understanding Errors from Multi-Input-Multi-Output (MIMO) Testing of a Cantilever Beam

Fernando Moreu, James Woodall, and Arup Maji

Abstract Multi-Input-Multi-Output (MIMO) testing is challenging due to the need to estimate inputs required to achieve desired responses at multiple locations on the test article of interest. Errors in the estimation of inputs can come from either the forward process (measurement of the Transfer Function (TRF) matrix that captures the input-output relationship) or the inverse process requiring the inversion of a rectangular matrix using the Moore-Penrose Pseudo-Inverse method.

In order to understand how small variation in the location of actuators contribute to errors, a set of SIMO (Single Input Multiple Output) experiments were conducted. A "Smartshaker" actuated a cantilever beam while sensors collected the acceleration response at two locations (close to the input location and at the tip of the cantilever beam). Researchers compared the response from different inputs with that estimated using the Transfer Function (TRF) to determine errors in the forward process. First, an applied Band-Limited-White-Noise (BLWN) signal informed the creation of the TRFs. Subsequently various narrow-band inputs at different frequencies and real-life vibration environments (earthquake and train) excited the structure. The study helps to understand the sources of uncertainty in MIMO vibration testing which can extend to more complex geometries and enables actuator placement that would minimize errors for different real-life inputs.

Keywords MIMO · Multi-input · Vibration · Testing · Error

1 Introduction

Researchers use Multiple Input-Multiple Output (MIMO) testing to take field response data and solve for the inputs to the system [1, 2]. This allows one to use response data which is readily available by the application of sensors to deduce what forces acted on a given structure or dynamic system. These techniques also apply to fields such as active feedback control for dynamic structures [3–5]. Researchers often use a simple beam to conduct MIMO analysis, both in the field and to test aspects of MIMO itself [6–9]. Due to the value offered by MIMO testing in dynamic testing and other engineering fields, there is a wealth of literature and study into the subject. Even so, MIMO still warrants further research and better understanding for many of its facets to better apply and learn from it, and researchers continue to develop new methods and applications for MIMO [10, 11].

One of the fundamental parts of common MIMO testing is the TRF matrix which derives the inputs. This quantity represents all relevant relationships for a test in dynamic analysis, relating all outputs to each of the inputs. Consequently, researchers have investigated the creation of this matrix, developing methods to improve accuracy and reliability when defining the system properties. There are many approaches to this step, depending on factors such as data availability, signal properties, and the conditioning of the TRF matrix [12]. In cases such as with larger structures, the application of a known

This paper describes objective technical results and analysis. Any subjective views or opinions that might be expressed in the paper do not necessarily represent the views of the U.S. Department of Energy or the United States Government.

F. Moreu (✉) · J. Woodall
Department of Civil, Construction, and Environmental Engineering, University of New Mexico, Albuquerque, NM, USA
e-mail: fmoreu@unm.edu; jameswoodall@unm.edu

A. Maji
Sandia National Laboratories, Albuquerque, NM, USA
e-mail: amaji@sandia.gov

K. Grimmelsman (ed.), *Dynamics of Civil Structures, Volume 2*, Conference Proceedings of the Society for Experimental Mechanics Series, https://doi.org/10.1007/978-3-030-77143-0_15

force to develop the needed relationships are infeasible, leading to much study in the field of output only modal identification [13–15]. Often, the ambient forces measured for these tests have relatively low magnitude and noise dominates their spectra, as studied by Wang [16], as well as Hazra [17].

MIMO inverts the completed TRF matrix to infer the input signals. In many ways, this is the key step which defines MIMO. Many factors can affect this step, and many researchers continue to investigate problems in the signal processing and the inversion process [18]. Thite [19] investigated how the output sensor locations impact the condition number of the Transfer Function (TRF) matrix, and how the locations may be optimized to minimize the condition number, reducing the error propagation in the inversion process.

Researchers use Finite Element Models (FEMs) in modal analysis and to perform MIMO in virtual experiments. However, when creating a model to represent a system, the presence of sensors on the real member can cause discrepancies between reality and the FEM. When these discrepancies are too severe, researchers must find ways to reconcile the systems [20]. One method of doing this is by mitigating the effect of the sensor on the real system [21–24]. Researchers also reconcile the systems by including the mass of the sensor within the model to match with an experiment. Once the model adequately matches the experiment, researchers may remove the sensor masses from the FEM to better simulate the behavior of the SUT in the field, when no sensors are present. However, many of these methods assume the location of the sensors to be known precisely, and the decoupling procedures can be prone to severe error propagation [25]. The dynamic effect of sensor mass is too pronounced to ignore in many tests, such as those in which the sensor mass is not insignificant compared to that of the structure under test. This paper investigates a case study with imprecise sensor location, and the consequential errors.

2 MISO Test Setup with Small Variation in Actuator Location

Figure 1 describes the SIMO experiments on a steel 34.5 × 1.5 × 0.125 in. cantilever beam with input actuator at 1/3 height and response accelerometers at 1/3 height and the tip. A load cell sensor between the stinger and the beam measured the force applied. Researchers conducted three sets of experiments with the accelerometer located at the 1/3 point and displaced ±7/16″ (1.1 cm) from the 1/3 point. The actuator was a Model K2007E01 "Smartshaker," manufactured by The Modal Shop Inc., with up to 7 pounds (31.2 N) peak sine force and a maximum stroke length of 1/2″ (1.3 cm). Acceleration sensors were Model 353B03 quartz shear accelerometers, manufactured by PCB Piezotronics. The force transducer was a Model 208C02 ICP® force sensor with a measurement range of 100lbf, also manufactured by PCB Piezotronics.

Researchers conducted the following sequence of experiments; the following section discusses the results:

1. Researchers applied a Band-Limited-White-Noise (BLWN) ranging from 0 to 250 Hz to the three experiment setups with the different sensor locations. The data from these tests informed the creation of the Transfer Functions (TRFs) at the 1/3 point and the tip, with the MatLab™ command *tfestimate*.
2. In the centered case, the stinger applied both narrow-band sinusoidal inputs and various real-life vibration inputs (earthquake, trains) (Fig. 2) to the beam. Researchers collected each of the responses as well as the force applied by the stinger.
3. For each test performed in step 2, researchers estimated the output responses by multiplying the measured inputs with each of the TRFs generated in step 1 (with conversions between time and frequency domain).
4. Researchers compared the measured outputs from step 2 with the estimated outputs from step 3, quantifying the discrepancy with the root mean square error (RMSE, or RMS error) between them. Equation (1) defines the equation for this metric.

$$\mathrm{RMSE} = 100 * \frac{\sqrt{\sum_{i=1}^{n} \frac{(\mathrm{Estimate}_i - \mathrm{Measured}_i)^2}{n}}}{\text{Measured amplitude}} \quad (1)$$

Researchers also performed the same series of steps with the sinusoidal inputs in a virtual simulation to examine the purely theoretical deviations caused by sensor displacement.

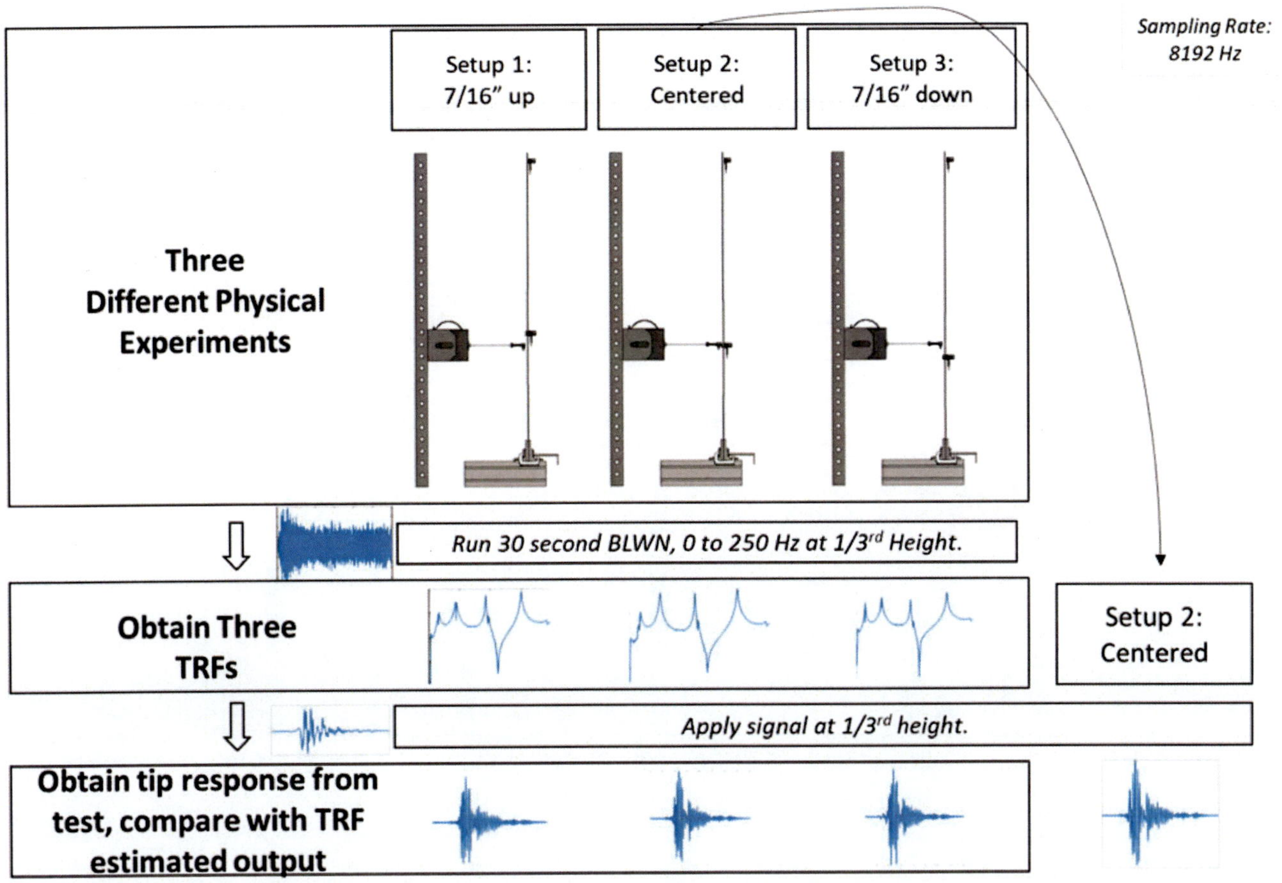

Fig. 1 Schematic of SIMO experiment

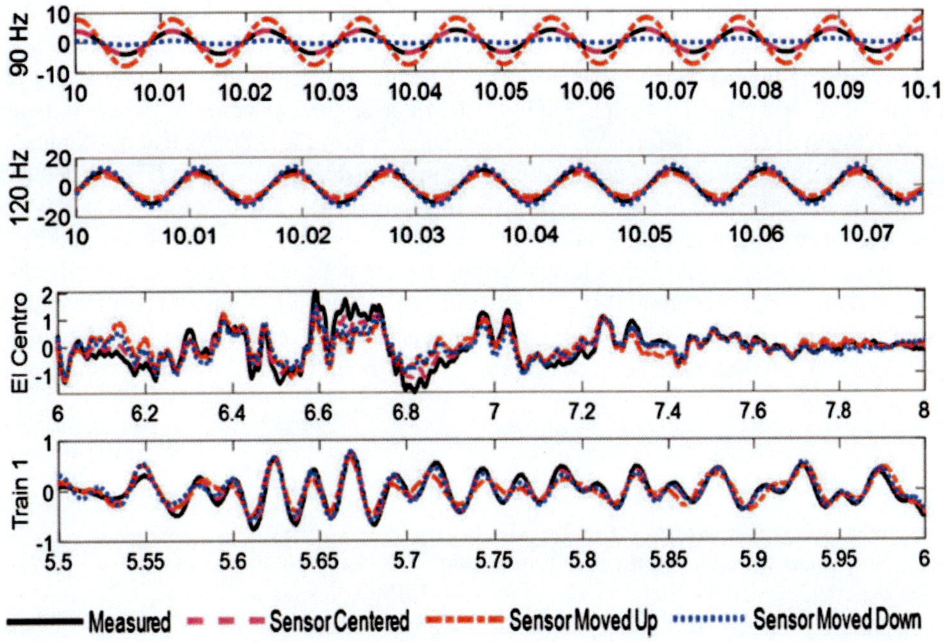

Fig. 2 Examples of time-domain data at 1/3 height for different inputs; measured and estimated

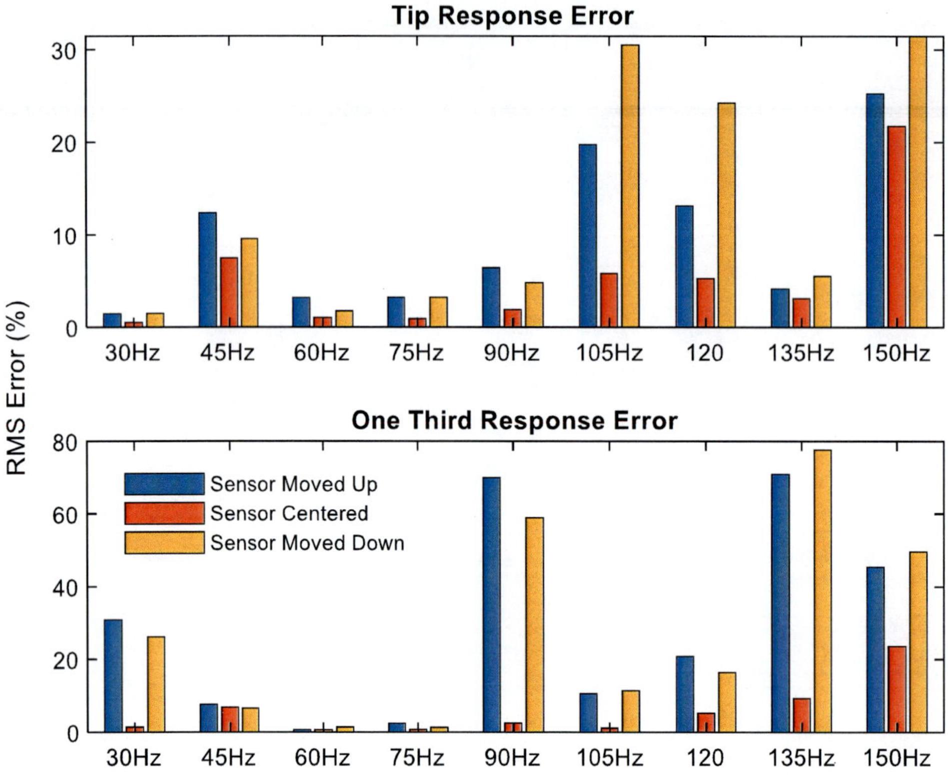

Fig. 3 Experimental RMS error for sinusoidal inputs

3 Test Results and Analyses

Figure 2 shows some example time-domain signals measured at the 1/3 height; the top two are narrow-band inputs at 90 and 120 Hz, followed by the El Centro earthquake and a train crossing a bridge. Figure 3 shows the RMS error between the estimated and measured time-domain signal described earlier for the sinusoidal inputs, while Fig. 4 shows the RMS error for the real-life vibration inputs. In Fig. 4 (right side), the RMS error corresponding to "Train 1" is lower than those corresponding to the BLWN and El Centro earthquake. A closer examination of Fig. 1 shows that the dominant frequency for Train 1 was $\approx$ 45 Hz, which had relatively small error for the narrow-band 45 Hz only input (Fig. 3). However, the BLWN had a higher error since it is an agglomeration of the error across all frequencies. The El-centro error was in-between since it had a dominant low-frequency content and some higher frequency (above 45 Hz) as well. All of the trains have similar frequency content, justifying their similar RMS error levels. Examining the sinusoidal inputs, there are similar trends in error for both the experimental (Fig. 3) and simulation (Fig. 5) cases. Of note is the increase in error for the experimental case at higher frequencies, which is not present in the simulation, while the moved cases remain of similar magnitude. This suggests the presence of some additional factor besides the standard properties used in the model that can dramatically affect results.

4 Conclusions

This paper investigated how small variations in the locations of actuators can affect the accuracy of MIMO testing. A SIMO test and associated analyses determined the accuracy of the forward MIMO process with sensor location uncertainty.

Researchers demonstrate the effects of a small deviation in sensor location on MIMO results. This emphasizes the importance of precise measurements and experimental consistency, as even small changes can result in large increases in error. The sensitivity (variation) of the TRFs (Transfer Functions) due the sensor location can explain errors in response away from the actuator (tip response). Finally, the cumulative effect on real-life loads (earthquake and train-induced vibration) is based on the frequency content of those events and the sensitivity of actuator location at the dominant frequencies. The study

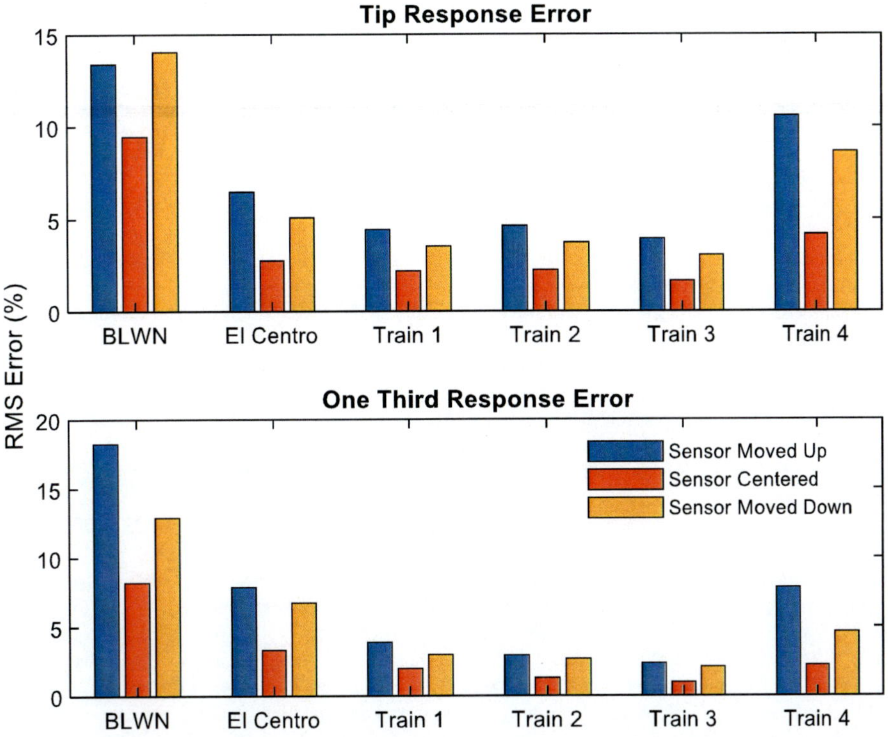

Fig. 4 Experimental RMS error for real-life inputs

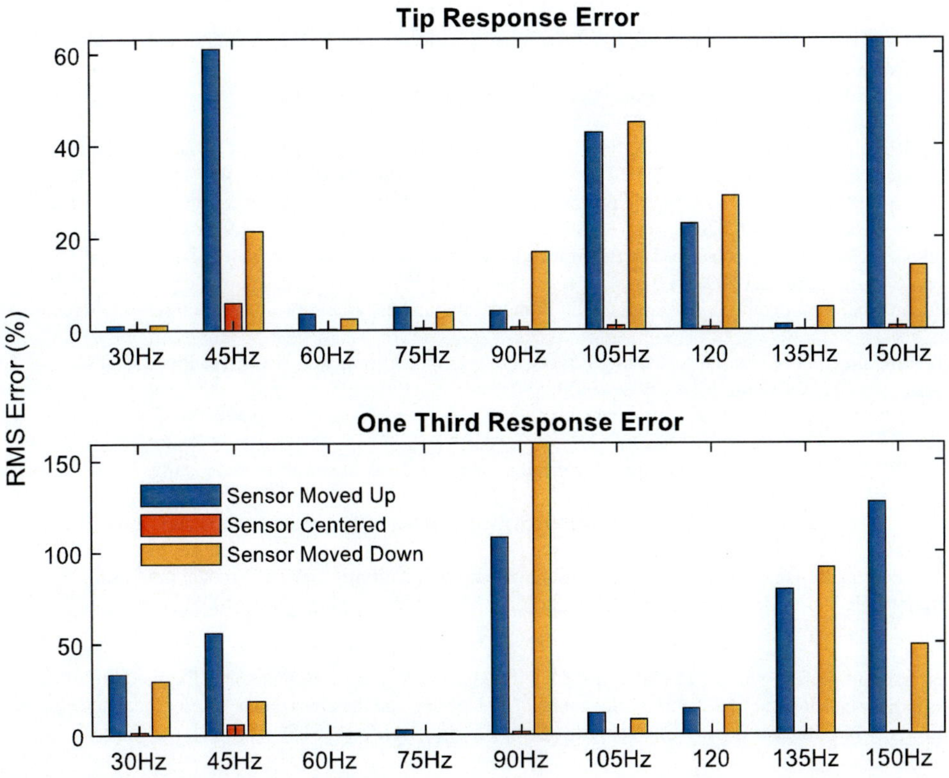

Fig. 5 Simulation RMS error for sinusoidal inputs

helps to understand the sources of uncertainty in MIMO vibration testing which can extend to more complex geometries and enables actuator placement minimizing errors for different real-life inputs.

References

1. Gombi, S.L., Ramakrishna, D.S.: Identification of impact force acting on a cantilever beam using inverse technique. In: Third International Conference on Smart Materials and Nanotechnology in Engineering (2012). https://doi.org/10.1117/12.921380
2. Carne, T., Mayes, R., Bateman, V.: Force Reconstruction Using the Sum of Weighted Accelerations Technique—Max-Flat Procedure. Sandia National Laboratories, Albuquerque (1993). https://doi.org/10.2172/10115211
3. Nestorović, T., Durrani, N., Trajkov, M.: Experimental model identification and vibration control of a smart cantilever beam using piezoelectric actuators and sensors. J. Electroceram. **29**(1), 42–55 (2012). https://doi.org/10.1007/s10832-012-9736-1
4. Saunders, W.R., Cole, D.G., Robertshaw, H.H.: Experiments in piezostructure modal analysis for MIMO feedback control. Smart Mater. Struct. **3**(2), 210–218 (1994). https://doi.org/10.1088/0964-1726/3/2/017
5. Tuma, J., Skutova, J.: Simulation of active vibration control of the cantilever beam. In: Proceedings of the 13th International Carpathian Control Conference (ICCC) (2012). https://doi.org/10.1109/carpathiancc.2012.6228745
6. Ratnam, C.: Dynamic response of a multiple cracked cantilever beam. Int. J. Res. Eng. Technol. **4**(12), 195–198 (2015). https://doi.org/10.15623/ijret.2015.0412038
7. Bateman, V.I., Mayes, R.L., Carne, T.G.: Comparison of force reconstruction methods for a lumped mass beam. Shock. Vib. **4**(4), 231–239 (1997). https://doi.org/10.1155/1997/642780
8. Mia, M.S., Islam, M.S., Ghosh, U.: Modal analysis of cracked cantilever beam by finite element simulation. Proc. Eng. **194**, 509–516 (2017). https://doi.org/10.1016/j.proeng.2017.08.178
9. Ma, C.-K., Lin, D.-C.: Input forces estimation of a cantilever beam. Inverse Prob. Eng. **8**(6), 511–528 (2000). https://doi.org/10.1080/174159700088027745
10. Smallwood, D.O.: Extreme inputs/outputs for multiple input multiple output linear systems. Shock Vib. (2005). https://doi.org/10.2172/875611
11. Houbolt, J.C.: A recurrence matrix solution for the dynamic response of elastic aircraft. J. Aeronaut. Sci. **17**(9), 540–550 (1950). https://doi.org/10.2514/8.1722
12. Mckelvey, T., Moheimani, S.R.: Estimation of phase constrained MIMO transfer functions with application to flexible structures with mixed collocated and non-collocated actuators and sensors. IFAC Proc. **38**(1), 219–224 (2005). https://doi.org/10.3182/20050703-6-cz-1902.00037
13. Yang, Y., Nagarajaiah, S.: Time-frequency blind source separation for output-only modal identification using independent component analysis. In: Safety, Reliability, Risk and Life-Cycle Performance of Structures and Infrastructures, pp. 1977–1983 (2014). https://doi.org/10.1201/b16387-288
14. Rainieri, C., Fabbrocino, G.: Output-only modal identification. In: Operational Modal Analysis of Civil Engineering Structures, pp. 103–210 (2014). https://doi.org/10.1007/978-1-4939-0767-0_4
15. Antoni, J., Garibaldi, L., Marchesiello, S., Sidhamed, M.: New separation techniques for output-only modal analysis. Shock Vib. **11**(3–4), 227–242 (2004). https://doi.org/10.1155/2004/560576
16. Wang, L., Wang, X., Li, X.: Inverse system method for dynamic loads identification via noisy measured dynamic responses. Eng. Comput. **33**(4), 1070–1094 (2016). https://doi.org/10.1108/ec-04-2015-0103
17. Hazra, B., Sadhu, A., Roffel, A.J., Narasimhan, S.: Hybrid time-frequency blind source separation towards ambient system identification of structures. Comput. Aided Civil Infrastruct. Eng. **27**(5), 314–332 (2011). https://doi.org/10.1111/j.1467-8667.2011.00732
18. Papanicolaou, V.G., Kravvaritis, D.: An inverse spectral problem for the Euler - Bernoulli equation for the vibrating beam. Inverse Prob. **13**(4), 1083–1092 (1997). https://doi.org/10.1088/0266-5611/13/4/013
19. Thite, A., Thompson, D.: Selection of response measurement locations to improve inverse force determination. Appl. Acoust. **67**(8), 797–818 (2006). https://doi.org/10.1016/j.apacoust.2006.01.0
20. Koruk, H.: Quantification and minimization of sensor effects on modal parameters of lightweight structures. J. Vibroeng. **16**(4), 1952–1963 (2014)
21. Decker, J., Witfeld, H.: Correction of transducer-loading effects in experimental modal analysis. In: Proceedings of the 13th International Modal Analysis Conference (IMAC13), pp. 1604–1608 (1995)
22. Cakar, O., Sanliturk, K.Y.: Elimination of transducer mass loading effects from frequency response functions. Mech. Syst. Signal Process. **19**(1), 87–104 (2005)
23. Bi, S., Ren, J., Wang, W., Zong, G.: Elimination of transducer mass loading effects in shaker modal testing. Mech. Syst. Signal Process. **38**, 265–275 (2013)
24. Ren, J., Wang, W., Bi, S.: Correction of transducer mass effect from the measured FRFs in hammer impact testing. Shock. Vib. **2017**, 6857326 (2017)
25. Van der Auweraer, H., Peeters, B., Donders, S.: Importance of uncertainty in identifying and using modal models. In: Proc. INCE Symposium on Managing Uncertainties in Noise Measurements and Prediction, Le Mans, France, June 27–29 (2005)

Fernando Moreu is an assistant professor in the Department of Civil, Construction and Environmental Engineering of the University of New Mexico, with courtesy appointments in the Department of Mechanical Engineering and the Department of Electrical and Computing Engineering. His PhD is from the University of Illinois at Urbana-Champaign in Structural Engineering (2015).

James Woodall is a graduate research assistant in the Department of Civil, Construction and Environmental Engineering of the University of New Mexico, pursuing a MS in structural engineering. His BS is from the University of New Mexico in the Department of Civil, Construction and Environmental Engineering (2018).

Arup Maji is a research scientist in Sandia National Laboratories. His PhD is from the University of Northwestern in Structural Engineering (1988).

Load-Displacement Behavior Clustering of RC Shear Walls Using Functional Data Analysis

Hamed Momeni and Arvin Ebrahimkhanlou

Abstract Reinforced concrete shear walls are one of the most common bracing systems in buildings. A significant number of existing buildings use RC shear walls designed for gravity loads and lateral loads. Several experimental studies have investigated the load-displacement behavior of these walls. As a standard approach, quasi-static cyclic loading is applied to the walls, and the load-displacement behavior (i.e. hysteresis loops) is captured. Due to the high number of points in these hysteresis loops, the comparison of the behavior of walls is complicated. This research takes advantage of functional data analysis (FDA) to analyze the hysteresis loops of walls. Functional principal component analysis (FPCA), as a dimension reduction tool, is used to reduce the dimensionality of hysteresis loops to just six dimensions. In this regard, test results of 189 different RC shear walls from 45 different research groups are used. Then, the K-means clustering method is applied to cluster the hysteresis loops based on their harmonic FPCA scores. The results show that just six dimensions are enough to represent the overall shape of hysteresis loops.

Keywords Functional data analysis · Functional PCA · Load-displacement behavior · Hysteresis loops · Concrete structures · Structural health monitoring · FPCA · High-dimensional data analytics · K-means clustering

1 Introduction

One of the most common bracing systems in buildings is reinforced concrete shear wall. Then, analysis and monitoring of such elements in the bulidngs are neccessay [26, 27]. The RC shear walls used in significant number of existing buildings are designed for gravity loads and lateral loads. Several experimental studies have investigated the behavior of these walls [1–5]. As a standard procedure, quasi-static cyclic loading is applied to the walls. Then, the loading displacement behavior, called hysteresis loops, is captured. These loops can be used to explain the behavior of walls and explain phenomena like strength degradation and pinching. However, there are standard load protocols; the walls are tested based on different loading setup with the various number of load steps and load rates. Furthermore, the test data were restored with a different number of data points. As mentioned, hysteresis loops data show full behavior of walls; analyzing these loops provides a comprehensive interpretation of load displacement of behavior walls. Due to the high number of dimensions that describes these loops, the comparison of the behavior of the walls needs complicated analysis. In this research, the key idea is to take advantage of functional data analysis and reduce the dimensionality of hysteresis loops to just two dimensions. Then, wall behaviors can be analyzed with less computational cost. The next section briefly reviews the theory of functional data analysis and then presents its implantation for evaluating hysteresis loops of RC shear walls.

1.1 Functional Data Analysis

Functional data analysis (FDA) deals with the analysis and theory of data that are in the form of functions [6]. Functional data are intrinsically infinite-dimensional. On one hand, the high intrinsic dimensionality of data constitutes challenges for theory and computation. On the other hand, the high or infinite-dimensional structure of data is a rich source of information.

H. Momeni (✉) · A. Ebrahimkhanlou
Department of Mechanical Engineering, New Mexico Tech, Socorro, NM, USA
e-mail: hamed.momeni@student.nmt.edu

K. Grimmelsman (ed.), *Dynamics of Civil Structures, Volume 2*, Conference Proceedings of the Society for Experimental Mechanics Series, https://doi.org/10.1007/978-3-030-77143-0_16

As a basic definition, functional data consist of a random sample of independent functions, $X_1(t), \ldots, X_n(t)$, on an interval $I = [0, T]$ on a line. These functions can be viewed as a one-dimensional stochastic process, which is defined as a collection of random variables indexed by elements of a set [7]. A common tool in FDA is functional principal component analysis that works as a dimension reduction tool. The next section presents this tool in detail.

1.2 Functional Principal Component Analysis

Principal component analysis [8] is a key dimension reduction tool for multivariate data. This traditional tool has been extended to functional data and is called functional principal component analysis (FPCA). FPCA works as a dimension reduction tool, and in sparse data situations, it can be used to impute missing functional data. FPCA facilitates the conversion of inherently infinite-dimensional functional data to a finite-dimensional vector of random scores. Similar to the PCA method, FPCA can express the underlying stochastic process by a countable sequence of uncorrelated random variables called the functional principal components or scores, which in many practical applications are truncated to a finite vector [7].

In the functional context, each principal component is specified by a principal component weight function $\xi(t)$, which is a vector defined over the range of t, the same as the range of functional data [9]. Equation (1) shows how principal component scores of object Y are calculated.

$$z_i = \int \xi_i(t) Y(t) dt \tag{1}$$

The aim of FPCA is to find the weight function $\xi(t)$ that maximizes the variance of the principal component scores z_i subject to the constraint; principal component weight functions should be an orthonormal basis. Equation (2) defines the normality constraints for weight function, and Eq. (3) defines orthogonality constraint.

$$\int \xi(t)^2 dt = 1 \tag{2}$$

The higher-order principal components are defined in the same way with the same constraints. In general, the jth component requires the constraints:

$$\int \xi_j(t)\xi_1(t)dt = \int \xi_j(t)\xi_2(t)dt = \int \xi_j(t)\xi_{j-1}(t)dt = 0 \tag{3}$$

As mentioned earlier, FPCA can facilitate dimension reduction as the first K terms, for large enough K, provide an approximation of data. Similar dimension reduction can be achieved by expanding the functional data into other function bases, such as spline, Fourier, or wavelet bases.

Several FDA applications are investigated in the field of structural health monitoring (SHM) and non-destructive evaluation (NDE). These applications are including model updating, correlation analysis, and missing data recovery. The next section reviews the recent studies that took advantage of the FDA in SHM and NDE.

1.3 Application of FDA in Structural Health Monitoring and Non-destructive Evaluation

In 2011, Zhou et al. used FDA to present a nonparametric degradation modeling framework for making inferences on signals that are observed sparsely or over short time intervals. They also used an empirical Bayes approach to update the stochastic parameters of the degradation model in real time using degradation signals for online monitoring. The application of this Bayesian framework is updating the residual lifetime of the monitored element of structure [10]. In 2014, a data recovery method was implemented by Sun et al. to perform fault diagnosis and prognosis in multichannel sensor systems. This method considers the possibly skewed distribution of each channel of signal trajectories, which can recover missing data for both individual and correlated sensor channels. This method also works with asynchronous data that may be sparse as well [11].

FPCA facilitates functional principal component regression by projecting functional predictors to their first few principal components, then employing regression models with vector predictors. Since FPCA is an essential dimension reduction tool, it is also useful for the classification and clustering of functional data. Li used FPCA for discrimination of time-series data. He combined FPCA with quantile-frequency analysis, which is a spectral analysis for transforming a time series into a bivariate function of quantile level and frequency variable. Using a case study on ultrasound signals, the proposed method shows that the new dimension-reduced features are useful for the non-destructive evaluation of aircraft panels [12].

Other categories of FDA applications take advantage of nonlinear methods like time warping. Nonlinear methods in the FDA, in addition to the common amplitude variation, consider time variation too. One particular application of the combination of FPCA and time warping is missing data imputation. In this regard, Chen et al. implemented an inter-sensor relationship modeling for strain monitoring. This modeling considers the temperature-induced strains as functional data. The phase and amplitude components were modeled by warping functions and a piecewise linear mapping model [13]. Not only nonlinear FPCA can recover the missing data, but also it can impute missing correlated probability density functions. As an example, a distribution regression approach has been implemented based on the warping transformation of distributions. Given a covariate PDF, the warping function is first estimated by a kernel regression model; then, the estimated warping function is used to transform the covariate PDF and obtain an imputation for the missing PDF [14]. Another example, an indirect distribution-to-distribution regression approach, has been proposed to restore distributions of missing SHM data via functional data analysis. In this particular application, the representation function of the missing distribution was first restored from a functional regression model [15, 16].

The functional data analysis technique is also used in SHM wireless network systems. In this regard, the FDA is used to analyze the correlation of probability distributions of inter-sensor monitoring data. The monitoring data collected by each sensor are divided into time segments and later summarized by the corresponding probability density functions (PDFs). The geometric relations of the PDFs in terms of their shape mappings between sensors are characterized by warping functions and decomposed into finite functional principal components [16].

2 Analysis

Several datasets were collocated from the result of experimental tests on RC shear walls [4, 17–20]. In this research, the test results of 189 different RC shear walls are used. These experiments are performed by 45 different research groups from 1977 to 2019. All of these walls are tested under quasi-static cyclic loads. However, they used different load setup and load protocols with different load steps and load rates. Figure 1 showed three different specimens with different load protocols and the different number of captured points. Table 1 summarized the experimental studies that are used in this research.

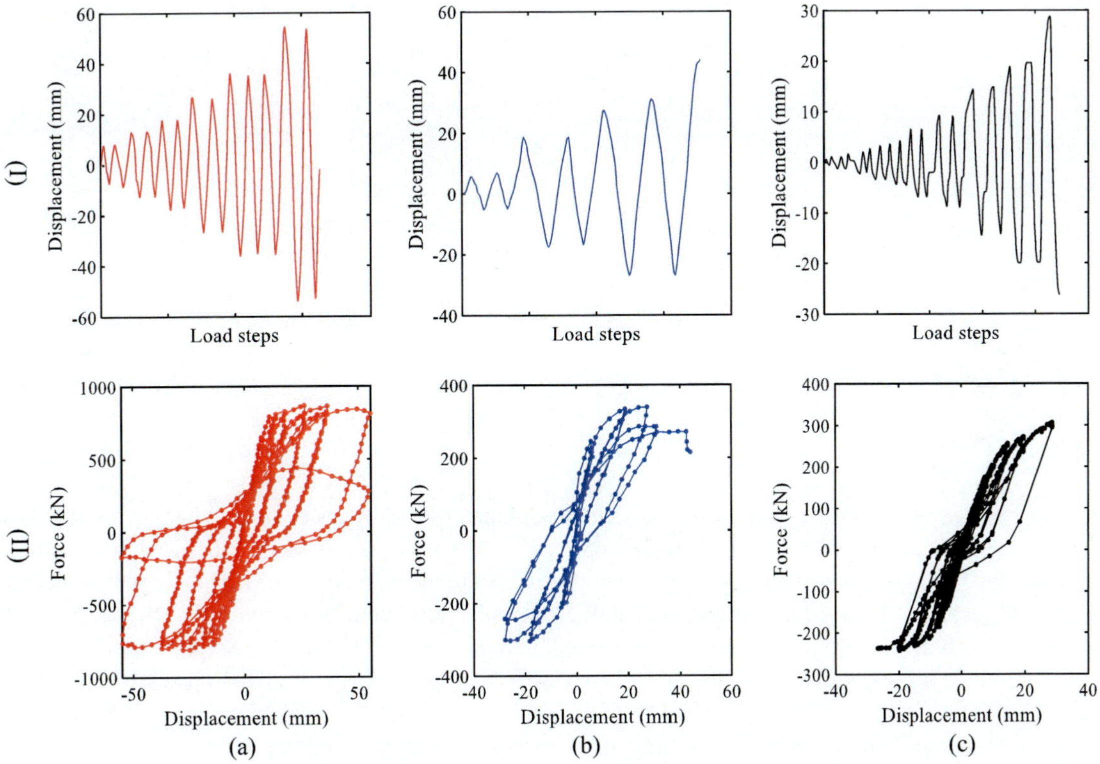

Fig. 1 (I) Applied load protocol and (II) hysteresis loops of three different specimens: (**a**) WSH1 [2], (**b**) RW-A15 [21], and (**c**) RCSW3 [22]

Table 1 Summary of experimental studies of RC shear walls SERIES database [17]

Research study	
Adajar et al. (1995)	Kimura and Sugano (1996)
Aoyama et al. (1984)	Layssi and Cook (2012)
Barda et al. (1977)	Matsuoka et al. (2003)
Belmouden and Lestuzzi (2006)	Nakachi et al. (1996)
Birely and Lehman (2008)	Nakamura et al. (2009)
Chen (2002)	Oh et al. (2006)
Chiou et al. (2004)	Oh et al. (2002)
Cho et al. (2004)	Palermo and Vecchio (2002)
Daniel et al. (1986)	Paterson and Mitchell (2003)
Dazio et al. (1999)	Paulay et al. (1982)
Deng, Liang, and Yang (2008)	Salonikios (1999)
Elnashai and Pinho (1998)	Shaingchin et al. (2006)
Esaki et al. (1994)	Sittipunt and Wood (2000)
Ghorbani-Renani et al. (2009)	Su and Wong (2006)
Greifenhagen and Lestuzzi (2006)	Takehara et al. (1997)
Hines et al. (2004)	Tasnimi (1998)
Hsiao et al. (2008)	Taylor et al. (1996)
Hwang et al. (2004)	Thomsen (1995)
Ji (2000)	Tokunaga and Nakachi (2012)
Jiang (1999)	Tran and Wallace (2012)
Kabeyasawa and Hiraishi (1998)	Wallace et al. (2008)
Kabeyasawa and Matsumoto (1992)	Zhang (2007)
Kabeyasawa et al. (1996)	

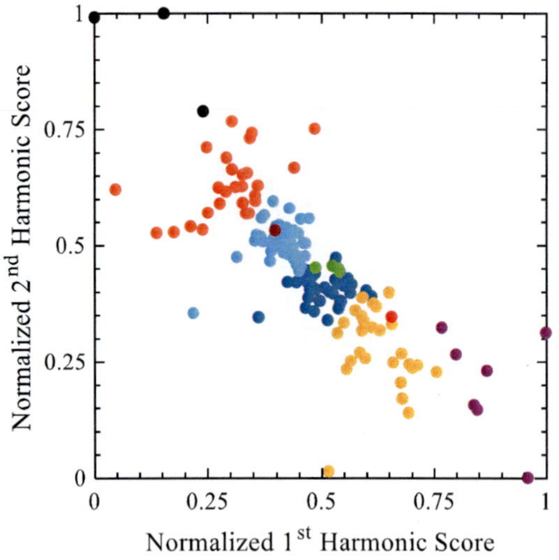

Fig. 2 First and second harmonic scores of hysteresis loops in *X*-direction

2.1 Analysis Results

First, for FPCA analysis, a splines basis of ten weight functions with the order of 4 was defined. After smoothing these basis functions, FPCA was applied on the dataset of hysteresis loops, and three harmonic scores are calculated in two directions of load and displacement. Then, the dimensionality of hysteresis loops was reduced to six, three harmonic scores in *X*-direction and three scores in *Y*-direction. Figure 2 shows the normalized first and second harmonic scores of each object.

Several clustering methods were implemented to group sets of similar concrete structures [23–25]. In this research, based on the obtained harmonic scores, a clustering method is applied to the hysteresis loops. This clustering is performed by the K-means method, and nine clusters are assumed. These clusters are shown with different colors in Fig. 2. Figure 3a shows the hysteresis loops assigned to each cluster. For some cases, the shape of hysteresis loops is not similar in a cluster;

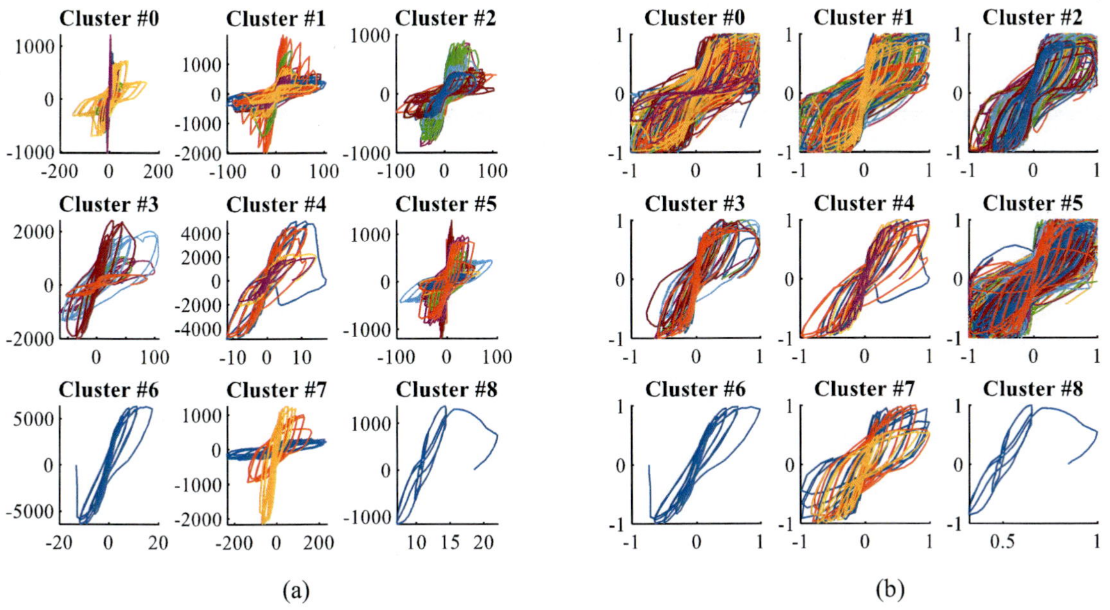

Fig. 3 The hysteresis loops in clusters (**a**) without normalization and (**b**) with normalization based on maximum force and displacement

however, by normalizing the hysteresis loops based on maximum absolute value force and maximum absolute displacement, this similarity can be presented, as shown in Fig. 3b.

3 Conclusion

In this study, the load-displacement behavior of 189 different RC shear walls is used. The wall specimens are tested under quasi-static cyclic loads. However, they used different load setup and load protocols with different load steps and load rates. Using FPCA, the dimensionality of hysteresis loops was reduced to six harmonic scores. Then, the K-means clustering method is used to cluster the hysteresis loops based on their harmonic FPCA scores. The results show that just six dimensions are enough to represent the overall shape of hysteresis loops. Further analysis can be performed to explain physical interpretation such as pinching in walls. Furthermore, a similar approach can be applied to other elements and structures.

It should be considered that there are limitations for the FDA. For example, FPCA methods are not robust against outliers because principal component analysis involves second-order moments. Outliers for functional data have many different facets that are due to the high dimensionality of data. They can appear as outlying measurements at a single or several time points or as an outlying shape of an entire function [7].

References

1. Paulay, T.: The design of ductile reinforced concrete structural walls for earthquake resistance. Earthquake Spectra. **2**(4), 783–823 (1986)
2. Dazio, A., Wenk, T., Bachmann, H.: Versuche an Stahlbetontragwänden unter zyklisch-statischer Einwirkung, vol. 239. ETH Zurich, Zurich (1999)
3. Tran, T.A., Wallace, J.W.: Experimental study of nonlinear flexural and shear deformations of reinforced concrete structural walls. In: 15th World Conf. Earthq. Eng. (2012)
4. Gulec, C.K.: Ultimate Shear Strength of Squat Rectangular Reinforced Concrete Walls. State University of New York, Buffalo (2005)
5. Basereh, S., Okumus, P., Aaleti, S.: Reinforced-concrete shear walls retrofitted using weakening and self-centering: numerical modeling. J. Struct. Eng. **146**(7), 4020122 (2020)
6. Ramsay, J.O., Silvermann, B.W.: Functional data analysis. Springer series in statistics. Biom. J. **40**(1), 56 (1998). https://doi.org/10.1002/(sici)1521-4036(199804)40:1<56::aid-bimj56>3.0.co;2-
7. Wang, J., Chiou, J., Mueller, H.-G.: Review of functional data analysis, pp. 1–47 (2015). http://arxiv.org/abs/1507.05135
8. Jolliffe, I.T.: Principal Component Analysis Springer Series in Statistics, XXIX, vol. 487, 2nd edn, p. 28. Springer, New York (2002)